THE WORLD OF ORGANIC AGRICULTURE

STATISTICS AND EMERGING TRENDS 2008

EDITED BY
HELGA WILLER, MINOU YUSSEFI-MENZLER AND NEIL SORENSEN

INTERNATIONAL FEDERATION OF
ORGANIC AGRICULTURE MOVEMENTS

Supported by

Schweizerische Eidgenossenschaft
Confédération suisse
Confederazione Svizzera
Confederaziun svizra

Swiss Confederation

Federal Department of Economic Affairs FDEA
State Secretariat for Economic Affairs SECO

First published by Earthscan in the UK and USA in 2008

For a full list of publications please contact:
Earthscan
2 Park Square, Milton Park, Abingdon, Oxfordshire OX14 4RN
711 Third Avenue, New York, NY 10017

First issued in paperback 2014

Earthscan is an imprint of the Taylor & Francis Group, an informa business

Notices
Practitioners and researchers must always rely on their own experience and knowledge in evaluating and using any information, methods, compounds,or experiments described herein. In using such information or methods they should be mindful of their own safety and the safety of others, including parties for whom they have a professional responsibility.

Product or corporate names may be trademarks or registered trademarks, and are used only for identification and explanation without intent to infringe.

All of the statements and results contained in this book have been compiled by the authors according to their best knowledge and have been scrupulously checked by the International Federation of Organic Agriculture Movements (IFOAM) and the Research Institute of Organic Agriculture (FiBL). However, the possibility of mistakes cannot be ruled out entirely. Therefore, the editors, authors and publishers are not subject to any obligation and make no guarantees whatsoever regarding any of the statements or results in this work; neither do they accept responsibility or liability for any possible mistakes, nor for any consequences of actions taken by readers based on statements or advice contained therein.

Authors are responsible for the content of their own articles. Their opinions do not necessarily always express the views of IFOAM or FiBL.

Additional information (links, graphs) is available at www.organic-world.net. Should corrections and updates of this book become necessary, they will be published at this site.

ISBN 13: 978-1-84407-592-8 (hbk)
ISBN 13: 978-1-138-01222-6 (pbk)

Willer, Helga, Minou Yussefi-Menzler and Neil Sorensen (Eds.) (2008)
The World of Organic Agriculture. Statistics and Emerging Trends 2008
International Federation of Organic Agriculture Movements (IFOAM) Bonn, Germany and Research Institute of Organic Agriculture (FiBL), Frick, Switzerland
ISBN IFOAM 978-3-934055-99-5
ISBN FiBL 978-3-03736-014-9

© 2008 IFOAM & FiBL. Published by Taylor & Francis.
International Federation of Organic Agriculture Movements (IFOAM) e.V., Charles-de-Gaulle-Str. 5, 53113 Bonn, Germany, Tel. +49 228 926 50-10, Fax +49 228 926 50-99, E-mail headoffice@ifoam.org, Internet www.ifoam.org. Trial Court Bonn, Association Register no. 8726, Executive Board: Gerald A. Herrmann, Alberto Lernoud, Mette Meldgaard

Research Institute of Organic Agriculture (FiBL), Ackerstrasse, 5070 Frick, Switzerland, Tel. +41 62 865 72 72; Fax +41 62 865 72 73, E-Mail info.suisse@fibl.org, Internet www.fibl.org

Hardback edition published by Earthscan, London, UK, 2008
ISBN Earthscan 978-1-84407-592-8 (hbk)
Language Editing: Neil Sorensen
Layout: Helga Willer, FiBL, Frick, Switzerland
Cover: Rob Watts
Graphs (if not otherwise stated): Claudia Kirchgraber, FiBL, Frick, Switzerland

A catalogue record for this book is available from the British Library.
Library of Congress Cataloguing-in-Publication data has been applied for.

TABLE OF CONTENTS

Alphabetical list of countries: Organic land, share of organic land, number of organic farms 2006
- Organic agricultural land by country 2006, sorted by importance - Share of organic agricultural land by country, sorted by importance - Organic land use by country 2006

Tables

Figures

Maps

Foreword Edition 2008

We herewith present the 2008 edition of the study 'The World of Organic Agriculture' documenting current statistics, recent developments and trends in global organic farming. The Research Institute of Organic Agriculture (FiBL), the International Federation of Organic Agriculture Movements (IFOAM) and the Foundation Ecology & Agriculture (SOEL) have been compiling the global statistical yearbook since 2000, with the support of NürnbergMesse. Since 2000, the latest global organic figures have been presented annually at the BioFach Fair in Nuremburg, of which IFOAM is the patron.

For this edition, the statistical information and all chapters have been updated. New additions include chapters on organic aquaculture, the FAO/IFOAM/UNCTAD International Task Force on Harmonization and Equivalence in Organic Agriculture (ITF) and a detailed report on organic farming in the Pacific Islands.

We are very thankful to the authors for contributing in depth information on their continent, their country or their field of expertise. We would furthermore like to express our gratitude to information and data providers as well as the editors, who have made the publication of this yearbook possible. We think that this product improves with each edition, and this is due to the huge commitment of numerous experts from all over the world.

Many thanks are due to Udo Funke of NürnbergMesse, the organizer of BioFach, who financially supported this as well as earlier editions of this study.

Furthermore, we have two new sponsors: the Swiss State Secretariat for Economic Affairs (SECO) / Economic Development and Cooperation (within the framework of its support activities for organic production in developing countries) and the International Trade Centre (ITC). We are very grateful to both for granting financial support for the global survey on organic farming.

Finally, we are very pleased to announce that this edition of 'The World of Organic Agriculture' is published in a hardcover edition by Earthscan (London, UK), and we are very grateful to Tim Hardwick at Earthscan for having initiated this cooperation. Earthscan is a leading publisher in environmentally sustainable development.

Bonn, Frick, Bad Duerkheim, February 2008

Angela Caudle de Freitas	Urs Niggli	Uli Zerger
IFOAM Executive Director	Director FiBL	Director SOEL
Bonn, Germany	Frick, Switzerland	Bad Duerkheim, Germany

Acknowledgements

Numerous individuals have contributed to the making of this book, as data and information providers, as authors or as supporters.

We are very grateful to all those listed below; without you it would not have been possible to produce this global statistical yearbook:

Acs Feketene, Gyorgyi, Control Union Certifications, Zwolle, Netherlands; **Adimado**, Samuel, Ghana Organic Agriculture Network GOAN; **Aiyelaagbe**, Issac, Organic Agriculture Project in Tertiary Institutions in **Nigeria**; **Akomagni**, Lazare, Sustainable Agriculture Development Network (RADAD), Cotonou, Benin; **Al Bitar**, Lina, Istituto Agronomico Mediterraneo IAMB, Valenzano, Italy; **Aldescu**, Teodora, Ministry of Agriculture, Organic Agriculture Department, Romania; **Alegría**, Beatriz, CLUSA, El Salvador; **Alföldi**, Thomas, Research Institute of Organic Agriculture (FiBL), Frick, Switzerland; **Altamirano**, Miguel, Instituto Interamericano de Cooperación para la Agricultura (IICA), Proyecto de Fomento a la Producción y Comercialización Orgánica de Nicaragua Cooperación Austriaca para el Desarrollo (ADA), Managua, Nicaragua; **Alvarado**, Fernando, Red de Agricultura Ecológica del Peru, Lima, Peru; **Alvarenga**, Fernando Rios, Servicio Nacional de Calidad y Sanidad Vegetal y de Semillas SENAVE, Asunción, Paraguay; **Amador**, Manuel CEDECO Corporación Educativa para el Desarrollo Costarricense, Costa Rica; **Apostolov**, Stoilko, Bioselena: Foundation for organic agriculture, Headoffice, Karlovo, Bulgaria; **Augstburger**, Franz, IMO Caribe, Santo Domingo, República Dominicana; **Babayev**, Vugar, Ganja Agribusiness Association, AZ-Ganja, Azerbaijan; **Barrios**, Lidia, SENASICA/SAGARPA, Mexico; **Ben Kheder**, Mohamed, Centre Technique de l'Agriculture Biologique, Sousse, Tunesia; **Bergleiter**, Stefan, Naturland e.V., Gräfelfing, Germany; **Bischof**, Andrea, Helvetas, Zürich, Switzerland; **Bouagnimbeck**, Hervé, International Federation of Organic Agriculture Movements, Bonn, Germany; **Büchel**, Klaus, Klaus Büchel Anstalt, Mauren, Liechtenstein; **Calderon**, Eduardo, Guatemala; **Carter**, Michelle, Care International Mozambique, Maputo; **Caudle de Freitas**, Angela B., International Federation of Organic Agriculture Movements IFOAM, Bonn, Germany; **Cierpka,** Thomas , International Federation of Organic Agriculture Movements IFOAM, Bonn, Germany; **Damghani,** Abdolmajid Mahdavi, Environmental Sciences Research Institute, Evin, Shahid Beheshti University, Evin, Tehran, Iran; **Darbinyan**, Nune, Ecoglobe - Organic control and certification body, Yerevan, Republic of Armenia; **Darolt**, Moacir, Instituto Agronomico do Parana, Curitiba Parana, Brasil; **de Braber**, Koen, Vietnam; **Eguillor Recabarren**, Pilar M., Ministerio de Agricultura, Oficina de Estudios y Políticas Agrarias, Santiago, Chile; **Ehmann**, Markus, HELVETAS Swiss Association for International Cooperation, BioCotton Project, Bishkek, Kyrgyzstan; **El-Araby**, Ahmed; Department of Soil Science, Ain Shams University-Cairo, Egypt; **Elola**, Sebastian, Movimiento Uruguay Organico, Uruguay; **Elvir Sanchez**, Sandra, Ministerio de Agricultura y Ganadería (MAG), Coordinadora de Agricultura Organica del Ministerio de Agricultura y Ganaderia, Tegucigalpa, Honduras; **Faber, Monique**, Administration des services techniques de l'agriculture (ASTA), Service de la protection des végétaux, Luxembourg; **Facknath**, Sunita, University of Mauritius, Faculty of Agriculture, Reduit, Mauritius; **Fernández Araya**, Claudia Andrea, AAOCH Agrupación de Agricultura Orgánica de Chile; **Fibiger Norfelt**, Tomas, Dansk Landbrugsrådgivning, Landscentret, Økologi, DK-Arhus Denmark; **Firmino**, Ana, Universidade Nova de Lisboa, Faculdade de Ciencias Sociais e Humanas, Lisboa, Portugal; **Fonseca**, Maria Fernanda, PESAGRO RIO. Rio de Janeiro, Brazil; **Funke**, Udo, BioFach - World Organic Trade Fair, Nürnberg Messe, Nürnberg, Germany; **García G**, Jaime E., Universidad Estatal a Distancia (UNED), Centro de Educación Ambiental (CEA), de Agricultura y Ambiente (AAA), San Pedro de Montes de Oca, Costa Rica; **Garibay**, Salvador, Research Institute of Organic Agriculture (FiBL), Frick, Switzerland; **Geber**, Ulrika, Swedish University of Agricultural Sciences SLU, Centre for Sustainable Agriculture CUL, Uppsala, Sweden; **Ghimire**, Maheswar, Kathmandu, Nepal; **Gómez Tovar**, Laura, Universidad Autónoma Chapingo, Mexico; **Gonzálvez Pérez**, Victor, Sociedad Española de Agricultura Ecologica

(SEAE) / Spanish Society for Organic Agriculture, Catarroja, Spain; **Gouri**, P.V.S.M., Agricultural and Processed Food Products. Export Development Authority, Ministry of Commerce and Industry, Govt. of India, New Delhi, India; **Guda**, Anula, SASA PIU, Swiss Development Cooperation, Albania; **Gunnarsson**, Gunnar Á., Vottunarstofan Tún ehf., Organic Inspection and certification, Reykjavik, Iceland; **Hadžiomerovic**, Maida, Organska Kontrola (OK), BA-Sarajevo. Bosnia & Herzegovina; **Harkaly**, Alexandre, Instituto Biodinâmico, Botucatu, Brazil, **Haumann**, Barbara, Organic Trade Association (OTA), USA-Greenfield, **Heinonen**, Sampsa, Finnish Food Safety Authority Evira, FIN-Loimaa; **Hensler**, Ines, Institut für Marktökologie IMO / Institute of Market Ecology, Weinfelden Switzerland; **Herrera Bernal**, Sandra Marcela, Cámara de Comercio de Bogota, Colombia; **Herrmann**, Gerald A., Organic Services, München, Germany; **Holmes**, Matthew, Organic Trade Association, Canadian Office, CAN-Sackville; **Holtmann**, Gabriele, International Federation of Organic Agriculture Movements (IFOAM), Bonn, Germany; **Huber**, Beate, Research Institute of Organic Agriculture, International Cooperation, Frick, Switzerland; **Iñiguez**, Felipe, EcoCuexco, Mexico; **Jasinska**, Aleksandra, Aberystwyth University; Jorjadze, Mariam, Elkana - Biological Farming Association, GE-Akhaltsikhe, Georgia; **Junovich**, Analía, SICA, Ecuador; Karoglan, Sonja, Ecologica, HR-Zagreb; **Kasterine**, Alexander, International Trade Centre (ITC), UNCTAD/WTO, Division of Product and Market Development, Geneva, Switzerland; **Katimbang-Limpin**, Lani, Organic Certification Center of the Philippines (OCCP), Barangay Laging Handa, Quezon City 1103, Philippines; **Kenny**, Lahcen, Institut Agronomique et Vétérinaire Hassan II, Horticulture & Agriculture Biologique, Agadir, Morocco; **Khodus**, Andrey, Agrosophie, RU-Solnechnogorsk; **Kilcher**, Lukas, Research Institute of Organic Agriculture (FiBL), International Cooperation, Frick, Switzerland; **Kirchgraber, Claudia**, Research Institute of Organic Agriculture (FiBL), Frick, Switzerland; **Kledal**, Paul Rye, University of Copenhagen Faculty of Life Sciences, Institute of Food and Resource Economics, Frederiksberg, Denmark; **Klingbacher**, Elisabeth, Bio Austria, Konsumenteninformation, A-Wien; **Kodikara**, Indumini, Sri Lanka Export Development, Colombo; **Koerber**, Hellmut von, Flexinfo, Ins, Switzerland; **Koesling**, Matthias, Norwegian Institute for Agricultural and Environmental Research, Bioforsk Organic Food and Farming Division, Tingvoll, Norway; **Kovács**, Annamária, Szent István University, Marketing Institute, Gödöllo, Hungary; **Kraokaw**, Sanayh, National Bureau of Agricultural Commodity and Food Standards, Bangkok, Thailand; **Kwai**, Noel, Tanzania Organic Agriculture Movement (TOAM), Dar es Salaam, Tanzania; **Larsen**, Poul H., Danmark Statistik - Agriculture and Transport, DK-Copenhagen; **Lehmann**, Sonja, GTZ Ecuador, Quito, Ecuador; **Lehocka**, Zuzana, Research Institute of Plant Production, Vyskumny ustav rastlinnej vyroby Piestany, SK-Piestany; **Lernoud**, Pipo, Buenos Aires, Argentina; **Liu**, Perrine, YU-SHI; **Lucas**, Amilcar, Care International, Mozambique, Maputo; **Lüthi**, Ruedi, Helvetas, PROFILE/PRORICE, Vientiane Capital, Laos; **Macey**, Anne, Canadian Organic Growers, CA-Saltspring Island, BC, Canada, V8K 2L6; **Mahmoudi**, Hossein, Environmental Sciences Research Institute, Evin, Shahid Beheshti University, Evin, Tehran, Iran; **Mandl**, Betty, MGAP DSGA, Uruguay; **Maohua**, Wang, China National Accreditation Administration of the People's Republic of China CNCA, Department for Registration, Certification and Accreditation Administration, China; **Mapusua**, Karen, Women in Business Development Inc, Apia Samoa; **Mason**, Seager, BIO-GRO New Zealand, Wellington; **Mc Auliffe**, Eddie, The Organic Farming Unit, Department of Agriculture, Food and Rural Development, Johnstown, Co. Wexford, Ireland; **Mendieta**, Oscar, Asociación de Organizaciones de Productores Ecológicos de Bolivia, La Paz; **Metera**, Dorota, Bioekspert, Jednostka Certyfikujaca w Rolnictwie Ekologicznym, RE-03/2003/PL, Warszawa, Poland, **Mikk**, Merit, Ökoloogiliste Tehnoloogiate Keskus, Centre for Ecological Engineering, EE-Tartu; **Milovanov**, Eugene, Organic Federation of Ukraine, UA-Kiev; **Miyoshi**, Satoko, IFOAM Japan, Saitama Japan; **Muga**, Juma, The Kenya Agriculture Organic Network (KOAN), Nairobi, Kenya; **Namuwoza**, Charity, National Organic Agricultural Movement of Uganda NOGAMU, Kampala, Uganda; **Niggli**, Urs, Research Institute of Organic Agriculture, CH-Frick; **O'Connor**, Bridget, Organic Producers & Processors Association of Zambia (OPPAZ), Lusaka, Zambia; **Omeira**, Nada, ALOA, -Beirut, Lebanon; **Ong**, Kung Wai, Grolink, Penang, Malaysia; **Oren Shnidor**, Pnina, Plant Protection and Inspection Services, Administration for Accreditation of Conformity Assessment Bodies, Standardization Department; Israel; **Padel**, Susanne, Aberystwyth University, Institute of Rural Sciences, UK-Aberystwyth SY23 3 AL Ceredigion; **Panyakul**, Vitoon, Green Net, Bangkok, Thailand; **Papastylianou**, Ioannis, Agricultural

Research Institute, Department of Farming Systems, Nicosia, Cyprus; **Parra**, Patricio, Chile; **Prawoto**, Agung, BIOCert, Bogor, Indonesia; **Ramos Santalla**, Nelson, Asociación de Organizaciones de Productores Ecológicos de Bolivia, La Paz; **Rajaonarison**, Andrianjaka, Laulanié Green Association (LGA), Antananarivo, Madagascar; **Randrianarisoa**, Sandra, Ecocert East Africa, Antananarivo, Madagascar; **Reynaud**, Michel, Ecocontrol GmbH, Northeim, Germany; **Richter**, Toralf, Bio Plus AG, CH-Seon; **Rios Alvarenga**, Fernando, SENAVE, Paraguay; **Rippin**, Markus, Agromilagro Research, Bornheim, germany; **Rison Alabert**, Nathalie, Agence bio, F-Montreuil-sous-Bois; **Robinson**, Dwight, Jamaica Organic Agriculture Movement (JOAM), Kingston 6, Jamaica W.I.; **Rodríguez**, Alda, Movimiento Uruguay Organico; **Rudmann**, Christine, Research Institute of Organic Agriculture (FiBL), Socio-Economics, CH-Frick; **Rundgren**, Gunnar, Grolink AB, S-Höje; **Sahota**, Amarjit, Organic Monitor Ltd., UK-London; **Salazar**, Julia, SENASA Perú, Responsable de la Autoridad Nacional Competente en Producción Orgánica del Perú; **Salem**, Sherif G., Department of Soil Science, Ain Shams University-Cairo, Egypt: **Sanborn**, Rebecca, Mayacert, Guatemala; **Sanchez**, Sandra Elvir, Ministerio de Agricultura y Ganaderia, Honduras, Coordinadora de Agricultura Organica; **Santiago de Abreu**, Lucimar, Sociologia - Embrapa Meio Ambiente, Jaguariúna, São Paulo, Brasil; **Schaack**, Diana, ZMP, Fachbereich Ökologischer Landbau, D-Bonn; **Scheewe**, Winfried, CEDAC, Phnom Penh; **Schmid**, Otto, FiBL, Research Institute of Organic Agriculture (FiBL), Frick, Switzerland; **Senic**, Iurie, Ministry of Agriculture and Food Industry, Department of Organic Agriculture and Plant Protection, Chisinau, Moldova; **Simon**, Stefan, EkoConnect - International Centre for Organic Agriculture of Central and Eastern Europe e.V., Dresden, Germany; **Slabe**, Anamarija, Institut za trajnostni razvoj, Institute for Sustainable Development, SI-Lubljana; **Smissen**, Nicolette van der, DIO, Alexandroupoli, Greece; **Sohn**, Sang Mok, Asian Research Network of Organic Agriculture (ARNOA), Dan Kook University, Research Institute of Organic Agriculture, Cheonan, Rep. of Korea; **Sorensen**, Neil; NGO Publishing; **Souhel,** Makhoul, General Commission for Scientific Agricultural Research, Damascus, Syria; **Süngü**, Erdal, Ministry of Agriculture and Rural Affairs, Horticulture Engineer, Ankara, Turkey; **Taylor**, Alistair, Agro Eco - Uganda Branch, Kampala, Uganda; **Taylor**, Gia Gaspard, Network of Non Governmental Organizations Trinidad and Tobago for the Advancement of Women, Director Information, Communications and Technology; **Trajkovic**, Radomir, Gradski dzid blok 4, PROBIO and Balkan Biocert, Skopje Macedonia; **Tas**, Petra, BIOFORUM Vlaanderen vzw, Beleidsmedewerker, Berchem; **Treffner**, Jens, Ekoconnect, Filderstadt, Germany; **Tshomo**, Kesang, Bio Bhutan, Thimpu, Buthan; **Twarog**, Sophia, UNCTAD, Trade, Environment & Development Branch, Geneva 10; **Ugas**, Roberto, Universidad Nacional Agraria La Molina, Peru; **Uwimbabazi**, Assinath, Rwanda Bureau of Standards, Kigali, Rwanda; **Vaclavik**, Tom, Green marketing, CZ-Moravské Knínice; **Vaheesan**, Saminathan, Helvetas Sri Lanka, Programme Officer/NRM, Colombo, Sri Lanka; **Wamba**, Guy, Ecocert, Doula, Cameroon; **Willer**, Helga, Research Institute of Organic Agriculture (FiBL), Frick, Switzerland; **Wong**, Jonathan, Hong Kong Organic Resource Centre, Agriculture, Fisheries and Conservation Department; **Wynen**, Els, Eco Landuse Systems, Flynn, Australia; **Yanogo**, Abdoul Aziz, Ecocert West Africa Office, Ouagadougou, Burkina Faso; Yombi Lazare, Ecoert, Douala, Cameroon; **Yussefi-Menzler**, Minou, Stiftung Ökologie & Landbau, Bad Dürkheim, Germany; **Zarina**, Livija, Priekuli Plant Breeding Institute, Agrotechnics Department, Priekuli, Latvia; **Zeballos**, Marta, Asociacion Rural del Urugua, Montevideo Uruguay; **Zenteno Wodehouse**, Virginia, Certificadora Chile Organico, Providencia, Santiago, Chile; **Zerger**, Uli, Stiftung Ökologie & Landbau, Bad Dürkheim, Germany, **Znaor**, Darko, Zagreb, Croatia.

11

Sponsors

We are very grateful our sponsors for granting financial support for of the 2008 edition of 'The World of Organic Agriculture' and the execution of the global data collection:

- Swiss State Secretariat for Economic Affairs (SECO)
  Economic Development and Cooperation (within the framework of its support activities for organic production in developing countries)
  Berne, Switzerland
  www.seco.admin.ch
- International Trade Centre (ITC)
  Geneva
  Switzerland
  www.intracen.org/dbms/organics
- NürnbergMesse, the organizers of the BioFach World Organic Trade Fair
  Nürnberg
  Germany
  www.biofach.de, www.nuernbergmesse.de

The World of Organic Agriculture 2008: Summary

Helga Willer[1], Neil Sorensen[2] and Minou Yussefi-Menzler[3]

Recent statistics

Organic agriculture is developing rapidly, and statistical information is now available from 138 countries of the world. Its share of agricultural land and farms continues to grow in many countries. According to the latest survey on organic farming worldwide, almost 30.4 million hectares are managed organically by more than 700'000 farms (2006). This constitutes 0.65 percent of the agricultural land of the countries covered by the survey (see chapter on the main results of the global organic survey 2008 and corresponding tables in the annex). In total, Oceania holds 42 percent of the world's organic land, followed by Europe (24 percent) and Latin America (16 percent). Currently (as of the end of 2006), the countries with the greatest organic areas are Australia (12.3 million hectares), China (2.3 million hectares), Argentina (2.2 million hectares) and the US (1.6 million hectares).

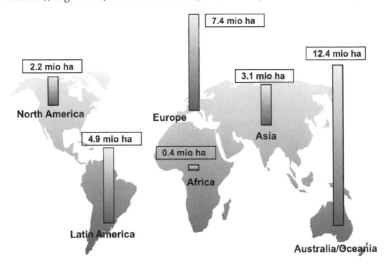

© SOEL, Source: FiBL Survey 2008

Map 1: Organic agriculture worldwide 2006

Source: FiBL Survey 2008, Graph: Minou Yussefi-Menzler, SOEL

---

[1] Dr. Helga Willer, Communication, Research Institute of Organic Agriculture (FiBL), Ackerstrasse, 5070 Frick, Internet www.fibl.org

[2] Neil Sorensen, NGO Publishing, www.ngopublishing.com

[3] Minou Yussefi-Menzler, Foundation Ecology & Agriculture (SÖL), Weinstrasse Süd 51, D-67098 Bad Dürkheim, Germany, www.soel.de

The proportion of organically compared to conventionally managed land, however, is highest in the countries of Europe. On a global level, the organic land area increased by almost 1.8 million hectares compared to the consolidated data from 2005 (see chapter on data consolidation, page 41). On all continents, the organic area has grown. The largest growth was in Oceania/Australia with 0.6 million hectares, followed by Europe where the organic area increased by half a million hectares and Asia with 0.4 million hectares.

The global survey on organic agriculture also contains information on the main land uses. At least some information on land use was available for more than 90 percent of organic land, showing that permanent grassland accounts for two thirds of the organic agricultural land and cropland for one quarter. In the context of the global survey on organic farming, data on certified organic wild collection are also collected. Thirty-three million hectares are certified for wild harvested products (2006). The majority of this land is in developing countries – quite the opposite of agricultural land, of which more than two thirds is in industrialized countries.

Market

Global demand for organic products remains robust, with sales increasing by over five billion US Dollars[1] a year. Organic Monitor estimates international sales to have reached 38.6 billion US Dollars in 2006, double that of 2000, when sales were at 18 billion US Dollars. Consumer demand for organic products is concentrated in North America and Europe; these two regions comprise 97% of global revenues. Asia, Latin America and Australasia are also important producers and exporters of organic foods. The global organic food industry has been experiencing acute supply shortages since 2005. Exceptionally high growth rates have led supply to tighten in almost every sector of the organic food industry: fruits, vegetables, beverages, cereals, grains, seeds, herbs, spices (see chapter on the global market by Amarjit Sahota).

Standards, regulations, and harmonization

The first standards on organic agriculture were developed by private organizations, and the IFOAM basic standards were first published in 1980 and have been continuously developed. In 2006 and 2007, the IFOAM World Board and Standards Committee organized two rounds of consultation on the draft version of the IFOAM Standards. These draft standards will be put to a vote after discussion at the General Assembly in June 2008 in Modena, Italy.

The need for clear and harmonized rules has not only been taken up by private bodies, IFOAM and state authorities, but also by United Nations Organizations. The FAO/WHO Codex Alimentarius Commission approved the Guidelines for the Production, Processing, Labeling and Marketing of Organically Produced Foods in June 1999, and animal production guidelines in July 2001. Currently the annex lists, which define what substances can be used in organic systems, are under revision, and in 2008 the discussion about a few controversial additives for food processing will continue.

---

[1] 1 US Dollar = 0.79703 Euros. Average exchange rate 2006.

The most important import markets for organic products are the European Union, the US and Japan, and thus their regulations have a significant impact on global trade and the development of standards in other regions. In 2007, the EU passed the first part of a completely revised regulation on organic production, which will come into force on January 1, 2009. The implementation rules referring to the revised regulation are expected to be published in mid 2008. Many European States that are not members of the EU will begin adapting their regulations to the EU Regulation in 2008/2009. Currently, 69 countries have implemented regulations on organic farming, and 21 countries are in the process of drafting a regulation (see chapter on standards and regulations by Beate Huber, Lukas Kilcher and Otto Schmid). In 2007, inter-agency collaboration in East Africa helped secure a major breakthrough in the adoption of the new East African Organic Products Standard. This standard was adopted by the East African Council of Ministers in April 2007 and launched together with the associated East African Organic Mark by the Prime Minister of Tanzania in May 2007 (see chapter by Sophia Twarog).

Today, 468 organizations worldwide offer organic certification services. Most certification bodies are in Europe (37 percent), followed by Asia (31 percent) and North America (18 percent). The countries with most certification bodies are the US, Japan, South Korea, China and Germany (see chapter on certification bodies by Gunnar Rundgren). Currently, 36 certification bodies have received IFOAM Accreditation by the International Organic Accreditation Services (IOAS), which assesses certification bodies against the IFOAM norms.

The International Task Force on Harmonization and Equivalence in Organic Agriculture (ITF), launched in 2003 and established jointly by FAO, IFOAM and UNCTAD, aims at achieving a general consensus on harmonizing private with government and government with government standards/regulations. The ITF is currently developing a set of essential International Requirements for Organic Certification Bodies as a basis for equivalence and future harmonization and a guidance document for judging equivalency of organic standards (Equitool) (see chapter by Sophia Twarog on the ITF).

Aquaculture

Certified organic aquaculture is a quite recent initiative. From the mid 1990s onwards, a number of certification agencies and organic growers' associations began developing specific aquaculture standards. In Germany and Austria, organic farmers who were running carp ponds as a source of additional income served as the main trigger for this process; they demanded adequate quality management and certification for this production. Since the mid 1990s, the progression of certified organic aquaculture has been characterized by a steady increase of product volumes on the market. Gradually, organic aquaculture lost its image as a purely niche activity, and bigger retail companies in Germany, the United Kingdom and Switzerland added aquaculture products into their assortment, which in turn encouraged more producers in many countries to convert to organic production.

Currently, the European Commission is drafting an organic aquaculture regulation, with the aim of regulating organic aquaculture in all EU countries. The US National Organic Program (NOP) is in a process comparable to that of the EU, and policies regulating organic aquaculture shall soon be integrated into the existing standard. Both processes are anticipated to be finalized by 2009 (see article by Stefan Bergleiter).

Africa

In Africa, there are more than 400'000 hectares of certified organic agricultural land. This constitutes about one percent of the world's organic agricultural land. There are at least 175'266 organic farmers. The countries with most organic land are Tunisia (154'793 Hectares), followed by Uganda (88'439 hectares) and South Africa (50'000 hectares). The highest shares of organic land are in Sao Tome and Prince (5.2 percent), Tunisia (1.6 percent) and Uganda (0.7 percent). It should be noted that some of the organic land in Africa that was previously classified as 'agricultural land' turned out to be certified wild collection, which plays a major role in Africa (more than eight million hectares). Important wild collection products are honey and gum arabic.

The majority of certified organic produce is destined for export markets, with the large majority being exported to the European Union. The African market for organic products is still small. Certified organic products are currently recognized in only a few domestic markets, including Egypt, South Africa, Uganda, Kenya and Tanzania. For exports, most African countries rely upon foreign standards. To date, the majority of organic production that is certified in Africa has been certified according to the EU regulation for organic products. Three countries have an organic regulation and seven are in the process of drafting one. Certification services are mainly offered by foreign-based certification bodies, some of which have established regional representation or developed closer cooperation with national bodies.

In many African countries, organic agriculture is not integrated into state agricultural policies. In some countries, mostly in East Africa, policy development is being undertaken, and the national organic movements are strongly involved in the process. A good example is the East African Organic Products Standard (EAOPS) and the associated East African Organic Mark (EAOM), which were developed by a public-private sector partnership in East Africa, supported by IFOAM, and the UNCTAD-UNEP Capacity Building Task Force on Trade. Both the mark and the standard were officially launched by the Prime Minister of Tanzania at the East African Organic Conference in May 2007 (see chapter by Hervé Bouagnimbeck on organic farming in Africa).

Asia

The total organic area in Asia is nearly 3.1 million hectares, managed by almost 130'000 farms. This constitutes ten percent of the world's organic agricultural land. The leading countries are China (2.3 million hectares), India (528'171 hectares) and Indonesia (41'431 hectares). The highest shares of organic land of all agricultural land are in Timor Leste (6.9 percent), Lebanon (1 percent), Sri Lanka and Israel (0.7 percent). Wild collection plays a major role in Azerbaijan, India and China (all with more than one million hectares).

The Asian market continues to show high growth in terms of organic food production and sales. Organic crops are grown across the continent, with some countries becoming international suppliers of organic commodities. Retail sales were about 780 million US Dollars in 2006 (Organic Monitor). Demand is concentrated in Japan, South Korea, Singapore, Taiwan and Hong Kong, the most affluent countries in the region.

As in other parts of the world, demand is surpassing supply with large volumes of organic foods imported into each country. The number organic trade fairs is also on the rise, a reflection of the growing demand.

Organic regulations have been established in eleven Asian countries, and another eight countries are in the process of drafting organic regulations. Organic regulations tend to be mandatory in importing countries and voluntary in exporting countries. Israel and India have attained equivalency status with the EU regulation on organic farming.

Organic farming increasingly receives government support, which is often reflected through the establishment of national regulations. Several countries, however, also have support programs for organic farming (India, Indonesia, Japan, Republic of Korea, Thailand, Vietnam) (see chapter on Asia by Ong Kung Wai).

Europe

Since the beginning of the 1990s, organic farming has rapidly developed in almost all European countries. As of the end of 2006, 7.4 million hectares in Europe were managed organically by more than 200'000 farms. In the European Union, 6.8 million hectares were under organic management, with almost 180'000 organic farms. 1.6 percent of the European agricultural area and 4 percent of the agricultural area in the EU is organic. Twenty-four percent of the world's organic land is in Europe.

The countries with the largest organic area are Italy (1'148'162 hectares), Spain (926'390 hectares) and Germany (825'539 hectares). The highest percentages are in Liechtenstein (29 percent), Austria (13 percent) and Switzerland (12 percent). Compared to 2005, the organic land increased by more than 0.5 million hectares. The increase is due to growth in most European countries, particularly in Spain, Italy, Poland and Portugal.

The biggest market for organic products in 2006 was Germany with a turnover of 4.6 billion Euros, followed by the UK (2.83 billion Euros). The highest market shares of organic products of the total market with around five percent are in Austria, Denmark and Switzerland. The highest per capita consumption of organic food is in Switzerland with more than 100 Euros spent on organic food per year and citizen. Some countries are currently experiencing a shortage of supply.

Support for organic farming in the European Union includes grants under the European Union's rural development programs, legal protection under the recently revised EU regulation on organic faming (since 1992) and the launch of the European Action Plan on Organic Food and Farming in June 2004. Countries that are not EU members have similar support. Furthermore, research into organic agriculture is supported both at a country as well as at an EU level, reaching at least 65 million Euros in 2006 (see the chapters on organic farming in Europe by Willer, Schaack, Padel et al. and Richter).

Latin America

In Latin America, 223'277 farms managed 4.9 million hectares of agricultural land organically in 2006. This constitutes 0.7 percent of the agricultural land in Latin America. Sixteen percent of the world's organic land is on this continent. The leading countries are Argentina (2'220'489 hectares), Uruguay (930'965 hectares) and Brazil (880'000 hectares). The highest shares of organic agricultural land are in Uruguay (6.1 percent), followed by Argentina (1.7 percent) and the Dominican Republic (1.3 percent).

Most organic production in Latin America is for export. From the coffee grains and bananas of Central America, to the sugar in Paraguay and the cereals and meat in Argentina, the trade of organic produce has been mostly oriented towards foreign markets. Countries like Argentina, Brazil and Chile have become important producers; however, over 90 percent of their organic crops are destined for export markets. Most organic food sales in these countries are from major cities like Buenos Aires and São Paulo.

Fifteen countries have legislation on organic farming, and three additional countries are currently developing organic regulations. Costa Rica and Argentina have both attained third country status according to EU regulation on organic farming.

Apart from regulatory support, many governments are now drafting action plans or similar programs. Furthermore, there is support for export promotion and for research (see article by Alberto Pipo Lernoud).

North America

In North America, almost 2.2 million hectares are managed organically, representing approximately a 0.6 percent share of the total agricultural area. Currently, the number of farms is 12'064. The major part of the organic land is in the US (1.6 million hectares in 2005). Seven percent of the world's organic agricultural land is in North America.

Valued at more than 17 billion US Dollars in 2006, the North American market accounted for 45 percent of global revenues. Growing consumer demand for healthy & nutritious foods and increasing distribution in conventional grocery channels are the major drivers of market growth (see chapter on organic farming in North America by Barbara Haumann).

In the United States, the National Organic Program has been in force since 2002. Canada has had a strong organic standard since 1999; this had been, however, voluntary and not supported by regulation. This will change in late 2008, when Canada's Organic Products Regulations are fully implemented by December 14. All products for sale in Canada will have to be certified to the Canadian standards and accredited by an accreditation body recognized by the Canada Organic Office (COO).

To help further domestic organic production, the organic sector during 2007 continued to encourage the US Congress to strengthen and support organic agriculture by incorporating provisions for organic farming in the 2007 Farm Bill. During 2008, the House and Senate will go to conference to resolve differences between the current two versions before a final bill is submitted to the White House for approval. In December 2007, the Canadian federal government committed over 1.2 million Canadian Dollars to the organic sector. Additional funds will be used to develop a national sector organization, the Organic Federation of Canada (OFC), to bring together all players in the industry, raise awareness of the sector and to help with regulatory development.

Oceania/Australia

This area includes Australia, New Zealand, and island states like Fiji, Papua New Guinea, Tonga and Vanuatu. Altogether, there are 7'594 farms, managing almost 12.4 million hectares. This constitutes 2.7 percent of the agricultural land in the area and 42 percent of the world's organic land. Ninety-nine percent of the land in the region is in Australia (12'294'290 hectares, 97 percent extensive grazing land), followed by New Zealand (63'883 hectares) and Vanuatu (8'996 hectares). The highest shares of all agricultural land are in Vanuatu (6.1 percent), Samoa (5.5 percent) and the Solomon Islands (3.1 percent). With an increase of 600'000 hectares in 2006, there has been a major growth of organic land in the region.

Growth in the organic industry in Australia, New Zealand and the Pacific Islands has been strongly influenced by rapidly growing overseas demand; domestic markets are, however, growing. In New Zealand, a key issue is lack of production to meet growing demand, for both export and the domestic markets.

Australia has had national standards for organic and biodynamic products in place since 1992, and like New Zealand, it is on the third country list of the European Union. It is expected that the Australian Standard, based on the National Standard employed since the early 1990s for the export market, will be adopted late 2008. In New Zealand, a National Organic Standard was launched in 2003.

There is little government support to encourage organic agriculture in Australia. However, over the recent past, governments have been supportive of the Australian Standards issue. In New Zealand, through the establishment of the sector umbrella organization Organics Aotearoa New Zealand (OANZ) and the Organic Advisory Programme (OAP), as well as other initiatives, there is political recognition of the benefits of organic agriculture (see chapters on Australia and New Zealand by Els Wynen and Seager Mason).

For the 2008 edition of 'The World of Organic Agriculture,' information on organic farming in the Pacific Islands was received for the first time. Organic agriculture shows considerable potential for development. The overall quantity of organic production and trading is still very small, even though some of the Pacific Islands have already reached relatively high shares of organic land; more than 20'000 hectares are under organic management. There is not yet a legal framework for organic agriculture. Although some countries have support programs, there is no an overall framework or cooperative strategy for development in the region (see chapter by Karen Mapusua).

Developments within IFOAM

Currently, global trade relations and rules, international and national policies, structural adjustments and trade concentration affect the self-image of the organic movement. In response, IFOAM strategically focused on issues relevant to the full diversity of the organic sector, including organic trade, food security, rural development, marketing of organic and regional products, the revision of the IFOAM Organic Guarantee System, the East Africa Organic Products Standard and the IFOAM recommendation on group certification.

In 2007, the FAO recognized organic agriculture as a significant and singularly effective alternative to conventional, chemical-based agriculture. IFOAM will continue to participate at FAO conferences in 2008, actively engaging IFOAM members and representatives. IFOAM will guarantee that the organic voice is heard at critical policy discussions world-wide, including at the United Nations Conference on Biodiversity and Planet Diversity (the World Congress on the Future of Food and Agriculture) to be held in Bonn in May 2008. (see chapter by Angela Caudle de Freitas and Gabriele Holtmann).

Organic Agriculture Worldwide: Current Statistics

HELGA WILLER[1]

1 Introduction

Since 1999, the Foundation Ecology & Agriculture (SOEL) and the Research Institute of Organic Agriculture (FiBL), in cooperation with the International Federation of Organic Agriculture Movements (IFOAM), have annually collected data about organic farming worldwide. This activity has been supported by NürnbergMesse, the organizers of BioFach Trade Fair since 1999. Since 2008 the work is also supported by the Swiss State Secretariat for Economic Affairs (SECO) / Economic Development and Cooperation (within the framework of its support activities for organic production in developing countries) and the International Trade Centre (ITC). For the 2008 edition, the global survey on organic agriculture was carried out for the ninth time (between September 2007 and January 2008).

The data – land area under organic management, land use and number of farms - show the current trends in organic agriculture worldwide, and can thus serve as an important tool for stakeholders, policy makers, authorities and consultants; they are useful in supporting strategies for organic agriculture as well as for monitoring the impact of support activities for organic agriculture.

For the ninth survey, 135 countries supplied data, showing the growing importance of organic agriculture.

2 Presentation of the statistics in this book

The statistics compiled in the survey can be found at various places in this book.

This chapter on the current statistics presents:

- Developments at a global and continent level, including organic land area, share of organic land, organic farms, and increase in organic land;
- Land use at a global level and by continent;
- Wild collection;
- Use of organic land on the continents; and
- Organic farming in developing countries.

In the ***continent chapters*** of this book, the following results of the global organic survey are available:

- Land area, share of total agricultural area and farms by country; and

---

[1] Helga Willer, Research Institute of Organic Agriculture (FiBL), Ackerstrasse, 5070 Frick, Switzerland, www.fibl.org

- Data sources for the country data. (These sources refer to all country related data, including land use and crop data.)

In the **annex**, the results of the global survey on organic farming are presented in full detail, including:

- Alphabetical country list with information on land under organic management, share of organic of agricultural land and numbers of farms;
- Country list with land under organic management, sorted by global importance;
- Country list with information on share of organic of agricultural land, sorted by global importance; and
- Country list with land use and crop details.

3 Scope of the survey

The global organic survey 2008 was carried out between September 2007 and January 2008.[1] Most of the data gained refer to the end of 2006.

Data on the total certified organic land, differentiated by agricultural use and wild collection, on land use details, on the number of farms, the value of the domestic market, as well as production volumes and exports, were surveyed. Whereas, as in previous years, a good picture on the total organic land and land use could be gained, the information on production and market volumes remains patchy, and a global picture cannot yet be drawn from these figures. For some countries, especially in Latin America, such production data are available. Some of these data are presented in the continent chapters.

Various information sources were used. Most of the data were supplied through national contact persons. We aimed at using the same contacts as for the previous survey. However, not all contacts responded, and as a result, new contacts had to be found for some countries. For a complete list of who provided data, see the tables in the continent chapters.

The contacts and information sources can be classified as follows:

- Member organizations of the International Federation of Organic Agriculture Movements (IFOAM);
- Contacts and data provided by staff of the Research Institute of Organic Agriculture (FiBL), of the Mediterranean Organic Agriculture Network (MOAN) and EkoConnect;
- National and international certification bodies; particular thanks are due to the Control Union, IMO, and Ecocert for providing data for the survey.
- Agricultural ministries;
- Eurostat, the Statistical Office of the European Union, Luxembourg, Data sets organic farming.

---

[1] Hervé Bouagnimbeck of IFOAM carried out the survey among the African countries and Helga Willer of FiBL among the other countries.

Countries and land use covered

In total, data on organic agriculture were available for 135 countries, and 124 countries provided new data (as of December 31, 2006). Thus, the survey covered approximately 70 percent of all countries (see table). This constitutes an increase compared to the 2007 survey, when data were available for a total 123 countries (Willer et al. 2007). For some countries, new information was only available for the total organic area, but updates on land use were unavailable, in which case the land use data of the previous survey were used. For some countries, only information on the agricultural area was available, but only the 2005 data for farms (and vice versa).

Table 1: Countries covered by the global organic survey

	Number of countries that provided new data (2006)	Total number of countries with data on organic land	Total countries	Percent of countries that provided data
Africa	28	30	55	55%
Asia	29	30	49	61%
Australia/Oceania	8	8	13	61%
Europe	41	42	44	93%
Latin America	17	23	33	70%
North America	2	2	02	100%
Total	124	135	196[1]	69%

- In Africa, data collection remains difficult. The availability and quality of information is improving, however, in many countries. With the exception of Tunisia, where the government collates the data, most of the data were supplied by private sector organizations. These are often umbrella organizations of the organic movement, who collect the data from the operators and certification bodies. In some cases, the data from only one certification body were available; the picture, therefore, often remains incomplete. For the following countries, data were supplied for the first time: Cameroon, Chad (wild collection), Congo, Ethiopia, Gambia, Ivory Coast, Sao Tome and Prince.

- Asia: More than 60 percent of the Asian countries answered the survey. Data availability is highly variable. In some countries, these data are supplied by government bodies (China, India) whereas in others, they are supplied by the private organic sector who collates the data from the certification bodies, operators or exporters. As a result, the picture remains incomplete for some countries. For the first time data for Mongolia were made available (wild collection).

- Europe: More than 90 percent of the European countries are covered by the survey. In Europe, the data availability is good, as most agricultural ministries collect and provide data on organic farming. Furthermore, the Eurostat database, which provides statistics for the member countries of the European Union, is very helpful.

---

[1] UN member countries, and Hong Kong, Niue, Palestine, Taiwan.

- Latin America: 70 percent of the countries in Latin America were covered. In South America, governments are increasingly providing detailed organic farming statistics, so the situation here has improved substantially since the first survey (Willer/Yussefi 2000). In Central America, the situation remains unsatisfactory for many countries; in most cases, the data are supplied by the organic sector, which often does not have access to the information of all certifiers or operators.

- North America: the United States and Canada supply very good data, including break-downs of land use patterns. For the US, the data are provided by the United States Department of Agriculture and for Canada by the Canadian Organic Growers.

- Oceania/Australia: For the first time, data were received for the Pacific Island States for this survey. For New Zealand, data are provided by the private sector, and for Australia by the governmental AQUIS (for this survey however, the data were supplied by the certifiers). For this continent, information on land use is limited.

Data storage

The data are stored in a database set up by flexinfo (www.flexinfo.ch). This database stores information on land area, land use, number of enterprises, and production and market data. The database is continuously and cooperatively developed with the German Central Market and Price Report Office (ZMP).

4 The main results of the global survey on organic agriculture 2008

4.1 Developments at a global and continent level

According to the FiBL Survey 2008, more than 30.4 million hectares were managed organically by more than 700'000 farms worldwide in 2006. This constitutes 0.65 percent of the agricultural land of the countries covered by the survey.

Table 2: Organic agricultural land and farms by continent 2006

Continent	Organic land area (hectares)	Share of total agricultural area	Organic farms
Africa	417'059	0.05%	175'266
Asia	3'090'924	0.17%	97'020
Europe	7'389'085	1.62%	203'523
Latin America	4'915'643	0.68%	223'277
North America	2'224'755	0.57%	12'064
Oceania	12'380'796	2.70%	7'594
Total*	30'418'261	0.65%	718'744

Source: FiBL Survey 2008

The continent with most organic agricultural land is Australia/Oceania, with almost 12.4 million hectares, followed by Europe with almost 7.4 million hectares, Latin America (4.9 million hectares), Asia (3.1 million hectares), North America (2.2 million hectares) and Africa (more than 0.4 million hectares).

As in previous years, Australia is the country with most organic land. China is second, followed by Argentina in third place. The ten countries with the most organic land have a combined total of almost 24 million hectares, constituting more than three quarters of the world's organic land. The annex includes a table with organic land in all countries, sorted by importance.

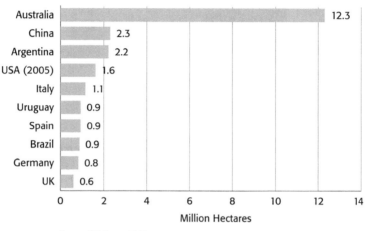

Source: FiBL Survey 2008

Figure 1: The ten countries with most organic land 2006

Source: FiBL Survey 2008

27

Distribution of global organic agricultural land by continent 2006

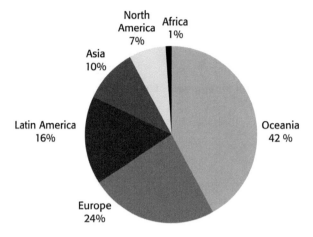

Source: FiBL Survey 2008

Figure 2: Distribution of global organic agricultural land by continent 2006

Source: FiBL Survey 2008

On a continent level, the share or organic land in proportion to all agricultural land is highest in Australia/Oceania (2.7 percent), followed by Europe. It should be noted, though, that some countries in Europe exhibit a much higher percentage; some countries have reached shares of more than ten percent of agricultural land (Austria, Switzerland). In the European Union, the share of organic land is four percent. In the annex, a table with all countries sorted by share of organic land is available.

In order to calculate the percentages, the data for most countries were taken from the FAO Statistical database FAOSTAT (as of 2003).[1] For the European Union, most data (as of 2005) were taken from Eurostat.[2] Where available, data for total agricultural land from ministries was employed (US, Switzerland, and Austria). Please note that in some cases the calculation of the shares of organic land and farms based on the Eurostat and FAOSTAT data might differ from the organic shares communicated by the ministries or experts.

---

[1] FAOSTAT, Data Archives, the FAO Homepage, FAO, Rome at faostat.fao.org > Data Archives > Land > Land Use; http://faostat.fao.org/site/418/DesktopDefault.aspx?PageID=418

[2] Eurostat, Agriculture & Fisheries Data, The Eurostat Homepage, Eurostat, Luxembourg, at ec.europa.eu/eurostat/ > Themes: Agriculture and Fisheries > Data > Agriculture, forestry and fisheries > Agriculture > Structure of agricultural holdings > Results of the farm structure surveys from 1990 onwards > General overview by area status > Key variables by region, agricultural area size classes and legal status; http://epp.eurostat.ec.europa.eu/portal/page?_pageid=0,1136206,0_45570467&_dad=portal&_schema=PORTAL

The ten countries with the highest shares of organic agricultural land 2006

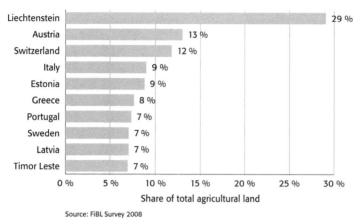

Figure 3: The ten countries with the highest shares of organic agricultural land 2006

Source: FiBL Survey 2008

Compared to the consolidated data of the previous survey (Willer et al. 2007), the organic land area increased by more than 1.8 million hectares (for details on data consolidation see separate section at the end of this chapter, page 41). Since the first global survey on organic agriculture (data mostly as of 1998; Willer/Yussefi 2000), the organic area has quadrupled.

Development of certified organic land worldwide 1998 to 2006

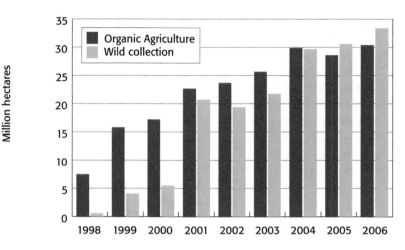

Source: FiBL, SOEL & IFOAM Surveys 2000 - 2008. Data consolidation in progress.

Figure 4: Development of certified organic land 1998 to 2006

Source: FiBL, SOEL and IFOAM: Surveys 2000 to 2008. The drop of organic land in 2005 was mainly due to the fact that large areas of extensive grassland went out of organic production in 2005, constituting a decrease of 1.2 million hectares in China, 0.6 million hectares in Chile, 0.3 million hectares in Australia).

The highest growth in 2006 was in Australia/Oceania, mainly due to a major increase of organic land in Australia. Furthermore, for the first time data had been supplied for the Pacific Islands (more than 20'000 hectares; see chapter by Mapusua in this book). In Europe, the organic land area increased by half a million hectares, in Asia by almost 0.4 million hectares and in Latin America by approximately 0.2 million hectares. The increase in North America is not as high, as new data were not yet available for the US. It is expected that the organic land area has increased substantially (see article by Barbara Haumann in this book).

Table 3: Organic agricultural land by continent 2005 (consolidated data)[1] and 2006 compared

Continent	2005 (hectares)	2006 (hectares)
Africa	331'697	417'059
Asia	2'682'630	3'090'924
Europe	6'886'078	7'389'085
Latin America	4'712'562	4'915'643
North America	2'199'225	2'224'755
Oceania	11'760'844	12'380'796
Total	28'573'037	30'418'261

Source: FiBL/SOEL surveys 2007 and 2008

In almost 90 countries, the organic land has increased since the previous survey. The highest increases were in Australia (almost 600'000 hectares), India (more than 300'000 hectares) and Uruguay (almost 200'000 hectares). For India in particular, this may be attributed to a growing domestic and export market. The ten countries with the highest increase of organic land had a growth of 1.6 million hectares.

Development of organic agriculture in the continents 2005 – 2006

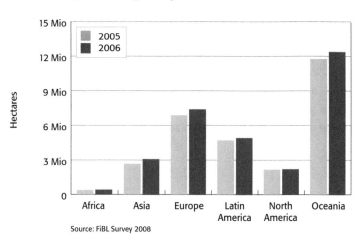

Source: FiBL Survey 2008

Figure 5: Development of organic agriculture 2005 to 2006 by continent

Source: FiBL/SOEL surveys 2007 and 2008

---

[1] For information on the consolidation of 2005 data, see separate section at the end of this chapter, page 41.

With almost 720'000 farms, considerably more farms were reported than for the previous survey (as of 2005, see Willer et al. 2007), especially in developing countries. According to the data obtained, the highest number of farms is in Mexico, followed by Uganda, Italy and India.

To find precise figures on the number organic farms remains difficult, as some countries report the number of smallholders, and others only the numbers of projects or grower groups. Some countries provide the number of producers per crop, and there may be overlaps for those growers who grow several crops. The global number of farms should consequently be treated with caution.

Organic farms by continent 2006

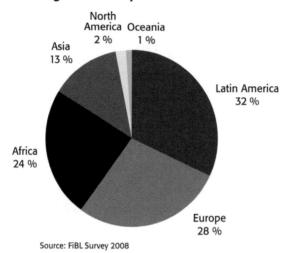

Source: FiBL Survey 2008

Figure 6: Distribution of organic farms by continent 2006

Source: FiBL Survey 2008

4.2 Land use

In this section, the results of the global land use survey are presented. When interpreting the data on land use in organic agriculture, it should always be kept in mind that detailed information on land use was not available for all countries. This means that the information presented is far from being complete, and with the available data, it has to be considered that the depth of information may differ[1] and that the data are aggregated in many statistics.[2]

For the data collected in the 2008 global survey, a slightly modified version of the classification system that was developed in 2006 (Baraibar 2006) was used. Details on the classification of the land use data are at the end of this chapter.

As in the previous survey (Willer et al. 2007), land use details were available for approximately 95 percent of organic land area. It should be noted, however, that this does not mean that detailed crop information is available for every country (see table in the annex on land use in all countries).

With a total of 4.45 million hectares, arable land accounts for one sixth of the organic agricultural area. Most of the world's organic arable land is in Europe, followed by North America and Asia. Most of the arable land is used for cereals, including rice, followed by field fodder crops.

Permanent crops account for five percent of the organic agricultural land (1.5 million hectares). Most of this land is in Europe, followed by Latin America and Africa. The most important crops are olives (almost a quarter of the permanent cropland) followed by coffee, fruits and nuts.

Permanent pastures/grassland (more than 20 million hectares) account for two third of the world's organic land. More than half of this grassland is in Australia. Furthermore, large areas of permanent pastures are in Latin America and Europe.

Looking at the land use at a continental level, a different pattern emerges for each continent. In the continent chapters, land use tables (main crop types) are available. Detailed information on land use patterns by country are available in the annex.

---

[1] For some countries, only information on the main uses (arable crops, permanent crops, and permanent grassland) was available. For Australia, for instance, only permanent grassland data were available. For other countries, very detailed statistical land use information can be found; the Danish statistics, for instance, list each vegetable type.
[2] Aggregation means that various crops are put together into one group. For instance, Spain combines cereals and leguminous crops, and it is thus impossible to have a figure solely for cereals. In such cases, the data available have to be classified as 'arable crops, no details.' In cases where arable and permanent crops were mixed, the category 'Cropland, no details' was used. As a result, a lot of information was lost due to the inability to obtain a precise breakdown of the data.

Table 4: Global organic agricultural land by main land use and crop categories

Main use	Main crop category	Agricultural area (ha)
Arable land	Arable crops, no details	359'459
	Cereals	1'722'617
	Fallow land as part of crop rotation	269'209
	Flowers and ornamental plants	366
	Green fodder from arable land	1'324'747
	Industrial crops	13'120
	Medicinal & aromatic plants	81'987
	Oilseeds	124'742
	Other arable crops	11'582
	Protein crops	211'913
	Root crops	33'304
	Seeds and seedlings	14'484
	Textile fibers	109'150
	Vegetables	178'014
Arable land total		*4'454'696*
Permanent crops	Citrus fruit	41'624
	Cocoa	93'308
	Coffee	340'722
	Fruit and nuts	323'425
	Grapes	107'426
	Industrial crops	83
	Medicinal & aromatic plants	7'375
	Nurseries	550
	Olives	381'337
	Other permanent crops	22'545
	Permanent crops, no details	25'428
	Sugarcane	24'285
	Tea	3'697
	Tropical fruit and nuts	98'985
Permanent crops total		*1'470'789*
Cropland, other/no details total		*1'523'480*
Permanent grassland	Cultivated grassland	74'778
	Pastures and meadows	3'890'681
	Permanent grassland, no details	15'539'844
	Rough Grazing	1'136'706
Permanent grassland total		*20'642'010*
Other	Aquaculture	4'222
	Fallow land	39'528
	Forest	191'114
	Unutilized land	40'071
Other total		*274'934*
No information		*2'052'352*
Total		30'418'261

Source: FiBL Survey 2008; Please note that information on land use, crop categories and crops was not available for all countries.

Table 5: Organic agricultural land by main main use and continent

Main use	Africa	Asia	Europe	Latin America	North America	Oceania/ Australia	World
Arable land	34'190	93'873	3'061'840	306'454	958'338		4'454'696
Permanent crops	163'447	66'126	701'103	494'692	45'321	100	1'470'789
Cropland, other/ no details	15'620	998'123	82'381	58'527		368'829	1'523'480
Permanent grassland	50'305	711'452	3'171'533	3'792'234	991'024	11'925'461	20'642'010
Other	150	1'034	269'850	3'900			274'934
No information	153'347	1'220'315	102'377	259'835	230'071	86'406	2'052'352
Total	417'059	3'090'924	7'389'085	4'915'643	2'224'755	12'380'796	30'418'261

Source: FiBL Survey 2008

Africa: For Africa (more than 400'000 hectares), information covering about half of the organic agricultural land was available. Most of this land is used for permanent crops. The main permanent crops are cash crops like olives (North Africa), coffee, cocoa, medicinal and aromatic plants.

Asia: Some land use details are known for two thirds of the organic land in Asia (3.1 million hectares). Arable land is mainly used for cereals, including rice. The most important permanent crops are coffee, fruits and nuts. Large areas of extensive grazing land are in China.

Europe: In Europe (7.4 million hectares), the organic land uses are relatively well known, and the main crop categories are well documented. Permanent pastures and arable land have approximately equal shares of the organic agricultural area. The main uses of the arable area are for cereals (1.2 million hectares), followed by the cultivation of field fodder (1 million hectares). Permanent crops account for nine percent of organic agricultural land. More than half of this land is used for olives, followed by fruits, nuts, and grapes.

Latin America: Most of the organic land in Latin America (4.9 million hectares), for which information was available, is permanent pasture. Permanent crops account for about eight percent of the agricultural area. The main crops are coffee, fruits, nuts and cocoa.

North America: In North America (2.2 million hectares), crop information was available for most of the land. Like in Europe, arable land and permanent grassland have almost equal shares. A major part of the arable land is used for cereal production.

Oceania/Australia: Most of the land in Australia is used for extensive grassland. Little or no information is available about the remaining land. Some land use details were available for New Zealand.

Land use in organic agriculture 2006 by continent

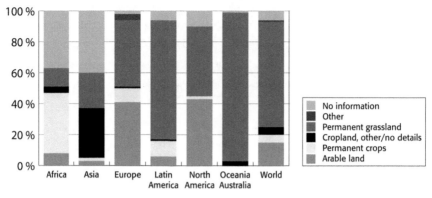

Source: FiBL Survey 2008

Figure 7: Land use in organic agriculture by continent 2006

Source: FiBL Survey 2008. Please note: information on land use, crop categories and crops was not available for all countries.

4.3 *Wild collection*

Certified organic wild collection is increasingly gaining importance. The collection of wild harvested crops is defined in the IFOAM Basic Standards (IFOAM 2006),[1] and wild collection activities are regulated in organic laws.[2]

---

[1] According to the IFOAM Basic Standards (2006):
- Wild harvested products shall only be certified organic if they are derived from a stable and sustainable growing environment. The people who harvest, gather, or wildcraft shall not take any products at a rate that exceeds the sustainable yield of the ecosystem, or threaten the existence of plant, fungal or animal species, including those not directly exploited.
- Operators shall harvest products only from a clearly defined area where prohibited substances have not been applied.
- The collection or harvest area shall be at an appropriate distance from conventional farming, pollution and contamination.
- The operator who manages the harvesting or gathering of common resource products shall be familiar with the defined collecting or harvesting area.
- Operators shall take measures to ensure that wild, sedentary aquatic species are collected only from areas where the water is not contaminated by substances prohibited in these standards.

[2] The recently revised EU regulation on organic production considers the collection of wild plants and parts thereof, growing naturally in natural areas, forests and agricultural areas as an organic production method - provided that those areas have not, for a period of at least three years before the collection, received treatment with products not allowed under the regulation. Furthermore, the collection must not affect the stability of the natural habitat or the maintenance of the species. The regulation also foresees standards for the collection of wild seaweeds and parts thereof.

Development of the organic wild collection area 2005-2006

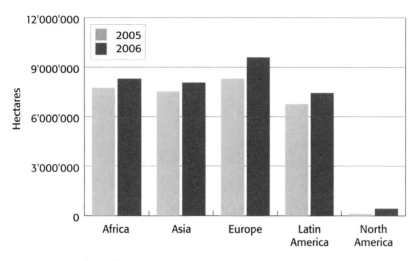

Source: FiBL Survey 2008

Figure 8: Development of organic wild collection area 2005 to 2006
Source: FiBL/SOEL Surveys 2005 and 2006

In 2005, a detailed survey on organic wild collection was carried out by Organic Services for the International Trade Centre (ITC). For the first time, in-depth information on the status of organic wild collection worldwide was made available. The study yields important information the extent of organic wild collection, cropping patterns and products (Censkowsky/Helberg 2007).

Within the scope of the FiBL and SÖL Surveys, wild collection land has been covered to some extent since 2003. For the 2008 survey, those providing data were asked to include basic figures on wild collection (land area). The data collected are not always consistent with the data gained within the frame of the ITC survey. As the results of the ITC survey were not updated, the results of the FiBL survey 2008 (data as of 2006) are presented in this edition of 'The World of Organic Agriculture.'

According to the FiBL Survey, more than 33 million hectares were certified for organic wild collection in 2006, constituting an increase of more than 3 million hectares compared to the 2005 data from FiBL. The highest increases between 2005 and 2006 were in Europe (mainly South Eastern Europe).

The organic wild collection area is more or less evenly distributed over four continents: Africa, Asia, Europe and Latin America, reflecting quite a different pattern than that for the agricultural land, most of which is in Oceania. There is almost no wild collection in Oceania/Australia and very little in North America.

Distribution of organic wild collection land by continent 2006

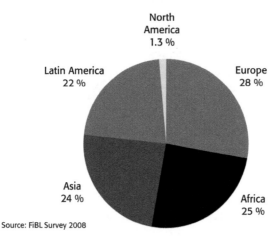

Source: FiBL Survey 2008

Figure 9: Organic wild collection land by continent 2006

Source: FiBL Survey 2008

The countries with the largest wild collection areas are Finland (mainly berries), followed by Zambia (bee pastures) and Brazil. Tables on the importance of organic wild collection in the countries are available in the continent chapters of this book.

Details on the collected crops were available (in hectares) for about half of the certified wild collection area. Fruit and berries play the most important role (mainly berries in Finland), followed by bee pastures (mostly in Africa) and by nuts (shea nuts in Africa and chestnuts in Latin America).

The ten countries with the largest wild collection areas 2006

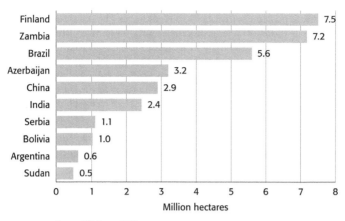

Source: FiBL Survey 2008

Figure 10: The ten countries with the largest certified wild collection areas 2006

Source: FiBL Survey 2008

Organic wild collection area by main type 2006

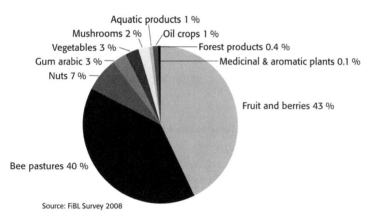

Source: FiBL Survey 2008

Figure 11: Main crop types in organic wild collection 2006. For about half of the wild collection area crop details were available.

Source: FiBL Survey 2008

39

4.4 Organic farming in developing countries

More than one quarter of the world's organic land is in developing countries (8.8 million hectares).[1] Most of this land is in Latin America followed by Asia, Africa and Europe. The leading countries in terms of organic land are China, Argentina, Uruguay and Brazil. The highest percentages of organic land are in several Pacific Island countries, East Timor, Uruguay and Argentina; in these countries, the shares of organic land of all agricultural land are comparable to those in Europe. These high shares can probably be attributed to a high potential for exports and to several support activities in these countries. Out of the developing countries covered by the survey, only few have a higher share of organic land than one percent of the agricultural area. Thus, compared to developed countries, organic farming lags behind.

The ten developing countries with most organic agricultural land 2006

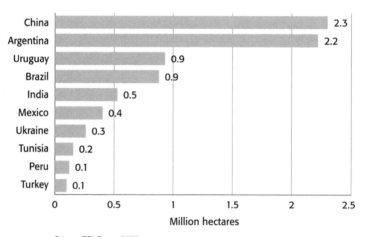

Source: FiBL Survey 2008

Figure 12: The ten developing countries with most organic land 2006

Source: FiBL Survey 2008; figures rounded

Even though land use details were not available for all developing countries, the statistics show that the shares of grassland (more than half of the organic land in these countries) and those of permanent crops are, compared to Europe and North America, relatively high. Arable land is of minor importance. This can be attributed to the fact that export plays an important role - either for meat products (mainly from Latin America) or for permanent crops. The most important permanent crops are export crops, such as coffee, olives, cocoa and sugarcane.

---

[1] For this section the countries listed on the List of Recipients of Official Development Assistance (ODA) of the Development Assistance Committee (DAC) of the Organization for Economic Cooperation and Development (OECD) were analyzed. The list is available at www.oecd.org/dataoecd/23/34/37954893.pdf

Use of organic agricultural land in developing countries 2006

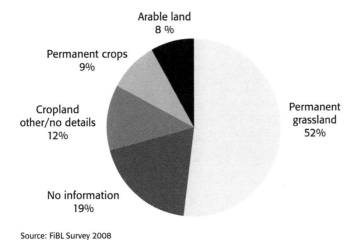

Source: FiBL Survey 2008

Figure 13: Land use in developing countries 2006

Source: FiBL Survey 2008

5 Consolidation of the 2005 data

For the 2008 global organic survey, it was possible to adjust some of the data gained in the previous survey (data as of 2005, Willer et al. 2007). These data are presented in the following table. For many countries, it has been possible to make a clear distinction between certified agricultural area and certified wild collection area. In many African and Latin American countries in particular, certified organic wild collection land had been classified as agricultural land in the previous surveys. For some countries, revised data have been made available by the authorities in charge.

Where the figures differ substantially from those communicated in the 2007 edition of 'The World of Organic Agriculture,' an explanation is given. Complete data sets are available at www.organic-world.net.

Table 6: Consolidated data on organic agricultural land as of 31.12.2005

Source: FiBL/SOEL Surveys 2007 and 2008; data adjustment process ongoing. Source: FiBL/SOEL Surveys 2007 and 2008

Country	Organic agricultural land 2005 (ha)	Country	Organic agricultural land 2005 (ha)
Albania	1'170	Italy	1'067'102
Algeria	887	Ivory Coast	0
Argentina	2'339'870	Jamaica	376
Armenia	265	Japan	8'109
Australia	11'715'744	Jordan	10
Austria	360'396	Kazakhstan	36'882
Azerbaijan	20'308	Kenya	2'274
Belgium	22'996	Korea, Republic of	6'095
Belize	1'810	Kyrgyzstan	221
Benin	284	Latvia	118'612
Bhutan	150	Lebanon	2'465
Bolivia	36'000	Liechtenstein	1'040
Bosnia Herzegovina	416	Lithuania	69'430
Brazil	842'000	Luxemburg	3'243
Bulgaria	14'320	Macedonia	249
Burkina Faso	332	Madagascar	2'220
Cambodia	952	Malawi	325
Canada	578'874	Malaysia	963
Chile	22'925	Mali	14'600
China	2'300'000	Malta	14
Colombia	45'647	Mauritius	175
Costa Rica	9'800	Mexico	307'692
Croatia	3'184	Moldova	11'075
Cuba (2005)	15'443	Morocco	6'812
Cyprus	1'698	Mozambique	716
Czech Rep.	254'982	Nepal	1'000
Denmark	147'482	Netherlands	48'765
Dominican Rep.	51'391	New Zealand	45'000
Ecuador	44'660	Nicaragua	51'057
Egypt	24'548	Niger	39
El Salvador	5'256	Nigeria	0
Estonia	59'862	Norway	43'033
Fiji	100	Pakistan	20'310
Finland	147'587	Palestine	1'000
France	560'838	Panama	5'244
Georgia	130	Paraguay	59'500
Germany	807'406	Peru	84'782
Ghana	17'261	Philippines	14'134
Greece	288'255	Poland	159'709
Guatemala	12'110	Portugal	212'728
Guyana	109	Romania	92'770
Honduras	1'823	Russian Federation, Asian Part	1'073
Hong Kong	12	Russian Federation, European Part	2'976
Hungary	128'574	Rwanda	105
Iceland	4'684	Saudi Arabia	13'730
India	150'790	Senegal	2'500
Indonesia	17'783	Serbia/Montenegro	591
Iran	0	Singapore	0
Ireland	35'266		
Israel	4'557		

Country	Organic agricultural land 2005 (ha)	Country	Organic agricultural land 2005 (ha)
Slovak Republic	90'206	Tunisia	143'099
Slovenia	23'499	Turkey	93'133
South Africa	50'000	Uganda	23'672
Spain	807'569	UK	619'852
Sri Lanka	10'049	Ukraine	241'980
Sweden	222'268	Uruguay	759'000
Switzerland	117'117	USA	1'620'351
Syria	20'500	Venezuela	16'000
Taiwan	1'441	Vietnam	6'475
Tanzania	38'875	Zambia	2'884
Thailand	21'701	Total	28'573'037
Timor Leste	21'526		
Togo	90		
Trinidad & Tobago	67		

- Argentina: In previous editions, the certified wild collection areas were listed under organic agricultural land.

- Bangladesh: The previously reported figure was not confirmed. Currently, according to Ministry sources, there is no certified organic land in Bangladesh. The figures reported previously probably referred to non-certified organic farming.

- Bolivia: In the previous editions, the wild collection areas were listed under organic agricultural land (see also Latin American chapter in this book).

- Colombia: For Colombia, no new data had been available for some years. The figures for the development of organic farming in Colombia have now been made available by the Ministry of Agriculture (see also Latin American chapter in this book).

- Chile: In previous editions, the certified wild collection areas were listed under organic agricultural land.

- Kenya: The figure reported in the previous years included certified wild collection.

- Korea: The data provided previously partly referred to environmentally friendly production.

- Morocco: In the previous editions, the area under certified wild collection (mainly argan oil) had been included.

- Poland: For Poland, the data as of 31.12.2005 communicated in the 2007 edition of the World of Organic Agriculture have been revised.

- Portugal: For Portugal, the data as of 31.12.2005 communicated in the 2007 edition of the World of Organic Agriculture have been revised by the authority in charge.

- Sudan: Previous data were not confirmed; most likely, much of the reported organic area was certified wild collection. Wild collection of gum arabic was confirmed by one certifier who reported 490'000 hectares for 2006.

- Uganda: The figure reported in the previous years included certified wild collection areas.

6 Background: Classification of land use data

For the data collected in the 2008 global survey, a slightly modified version of the classification system that was developed in 2006 (Baraibar 2006) was used. As the data were collected, a classification system was developed according to the kind of data received. FiBL and SOEL are planning to improve the classification system and to ultimately bring it in line with international classification systems. As with the 2006 survey (Baraibar 2006), the following problems were found:

- Standardization on a global level is lacking, and it is often difficult to compare the data, even though availability and quality of the statistics have improved in many countries.
- Aggregation levels (i.e. groupings of crops) vary significantly.
- Other ranking problems occurred when trying to classify a crop used in differing ways around the world (e.g. flax can be an industrial crop used for fiber or an oilseed).

For this survey, the general FAO classification[1] for main land uses[2] was utilized, with slight modifications. The following main levels were used to classify the land use and crop data: arable land,[3] permanent crops,[4] other cropland,[5] permanent grassland,[6] and other.[7] Furthermore, information on wild collection crops was included into the general survey, and the data were stored when available. 'No information' was used as a category for land where no details were available at all.

The information gathered was entered into a database created for this purpose. The data were entered into this database at three levels:

1) Main category (arable land, permanent crops, permanent pastures/grassland, Cropland without details, other land, wild collection).
2) Crop category (main crop groups like cereals)
   This second category was used to classify the main groups of crops within each main category. As the information provided was very different from one country to another, this classification level aims to include the most important crop groups.
3) Crop (individual crops)
   This last category includes specific crops grown organically. They can be as specific as 'maize for silage' and general as 'greenhouse cultivated vegetables.'

For the classification of arable crops, a classification very close to that of Eurostat, the European Statistical Office was used.

---

[1] See FAOSTAT homepage at http://faostat.fao.org/ > Metadata > Concepts and definitions > Methodology (search) > 'Land use'http://faostat.fao.org/site/362/DesktopDefault.aspx?PageID=362, Download of January 18, 2008.

[2] This system is used for instance by Faostat to classify land use data and can be found at FAOSTAT (http://faostat.fao.org/) > Archives > Land use and irrigation http://faostat.fao.org/site/418/default.aspx, download of January 18, 2008.

[3] According to the FAO arable land is land under temporary crops, temporary meadows for mowing or pasture, land under market and kitchen gardens and temporarily fallow land (less than five years). Abandoned land resulting from shifting cultivation is not included in this category. Data for 'Arable land' are not meant to indicate the amount of land that is potentially cultivable.

[4] According to FAO, permanent cropland is land cultivated with crops that occupy the land for long periods and need not to be replanted after each harvest, such as cocoa, coffee and rubber. This category includes land under flowering shrubs, fruit trees, nut trees and vines, but excludes land with trees grown for wood or timber.

[5] The FAO category 'Non-arable and permanent crops' was used for crops that did not fit into the other categories or for cropland for which details were not known, e.g. when crops of the arable and the permanent crop category had been put into one group (e.g. olives and annual oil crops).

[6] According to FAO, permanent pasture is land used permanently (five years or more) for herbaceous forage crops, either cultivated or growing wild (wild prairie or grazing land).

[7] The FAO has a category 'Forest and Woodland.' In the FiBL survey, forest, aquaculture and the unutilized land categories were all grouped under 'Other.'

- Cereals (Eurostat code C_01)
- Protein crops / Dried Pulses (Eurostat code C_02)
- Root crops (Eurostat code C_03)
- Industrial crops (Eurostat code C_04)
- Oilseeds (at Eurostat a subcategory of 'industrial crops' Eurostat code C_041)
- Textile Crops (at Eurostat a subcategory of 'industrial crops' Eurostat code C_044)
- Medicinal and aromatic plants and spices (in Eurostat a subcategory of 'other industrial crops' Eurostat code C_0451)
- Vegetables (Eurostat code C_05)
- Green fodder from arable land (Eurostat code C_06)
- Flowers and ornamental plants (at Eurostat a subcategory of 'other arable crops,' Eurostat code C_071)
- Seeds and seedlings (at Eurostat a subcategory of 'other arable crops,' Eurostat code C_072)
- Fallow land as part of the crop rotation (Eurostat code C_15)

For the classification of the permanent crops a classification very close to that of Eurostat, the European Statistical Office was used.

- Fruit and nuts (Eurostat code C_09; at Eurostat 'Fruit and Berries')
- Citrus fruit (Eurostat code C_10)
- Grapes (Eurostat code C_11)
- Olives (Eurostat code C_12)
- Coffee (No Eurostat Code)
- Cocoa (No Eurostat Code)
- Sugarcane (No Eurostat Code)
- Tropical fruit (No Eurostat Code)
- Tea (No Eurostat Code)

Cropland, other/no details includes permanent crops and arable crops together.

Permanent pastures include:

- Pastures and meadows (Eurostat Code c_081)
- Rough grazing (Eurostat Code c_082)

Other includes unutilized land, managed forest, aquaculture

For wild collection, a classification system still needs to be developed.

7 Survey 2009

The next global organic survey will start by mid 2008. We would be very grateful if data could be sent to us, but we will of course also contact all experts. Should you notice any errors regarding the statistical data in this volume, please let us know; we will then correct the information in our database and provide the corrected data in the 2009 edition. Corrections will also be posted at www.organic-world.net. The previous editions of 'The World of Organic Agriculture' can be downloaded here. Contact: helga.willer@fibl.org.

8 Further reading

Baraibar, Bàrbara (2006) Organic Coffee. In Willer/Yussefi 2006, Statistics and Emerging Trends. IFOAM, Bonn, Germany. pp 52-54, http://orgprints.org/5161/

Baraibar, Bàrbara (2006) Organic Cotton. In Willer/Yussefi 2006, Statistics and Emerging Trends. IFOAM, Bonn, Germany. pp 55-57, http://orgprints.org/5161/

Baraibar, Bàrbara (2007): Data Collection and Processing. In: Willer/Yussefi (Eds.) The World of Organic Agriculture 2006, International Federation of Organic Agriculture Movements, Bonn, Germany and Research Institute of Organic Agriculture (FiBL), Frick, Switzerland

Censkowsky Udo and Ulrich Helberg: Organic Wild Collection. In: Willer/Yussefi (eds.) 2007: The World of Organic Agriculture. Bonn/Frick 2007. Available at http://orgprints.org/10506/

Garibay, Salvador (2006) Organic Cocoa Production. In Willer/Yussefi 2006, Statistics and Emerging Trends. IFOAM, Bonn, Germany. pp 58-61, http://orgprints.org/5161/

Geier, Bernward (2006) Organic grapes – More Than Wine and Statistics. In Willer/Yussefi 2006, Statistics and Emerging Trends. IFOAM, Bonn, Germany. pp 62-65, http://orgprints.org/4858/

IFOAM (2006): The IFOAM Basic Standards for Organic Production and Processing. Version 2005. IFOAM, Bonn

Mayo, Robert (2004): Organic agricultural statistics and information at the United Nations Food and Agriculture Organization: initiatives, opportunities and challenges. In: Recke, Guido; Willer, Helga; Lampkin, Nic and Vaughan, Alison, Eds. (2004) Development of a European Information System for Organic Markets - Improving the Scope and Quality of Statistical Data. Proceedings of the 1st EISfOM European Seminar, held in Berlin, Germany, 26-27 April 2004. FiBL-Report. Research Institute of Organic Agriculture (FiBL), CH-Frick. http://orgprints.org/2935/

Rippin, Markus; Vitulano, Susanna; Zanoli, Raffaele and Lampkin, Nicolas (2006) Synthesis and final recommendations on the development of a European Information System for Organic Markets. = Deliverable D6 of the European Project EISfOM QLK5-2002-02400. Report, Institute of Rural Sciences, University of Wales. http://orgprints.org/8961/

Rippin, Markus; Willer, Helga; Lampkin, Nicolas and Vaughan, Alison, Eds. (2006) Towards a European Framework for Organic Market Information. Proceedings of the Second EISfOM European Seminar, Brussels, November 10 & 11, 2005. Research Institute of Organic Agriculture (FiBL), Frick, Switzerland. http://orgprints.org/6054/

Willer, Helga and Minou Yussefi (2000): Organic Agriculture Worldwide. Statistics and Future Prospects. Bad Dürkheim, Germany.

Willer, Helga, Minou Yussefi and Dirk Sthamer (2007): The Global Survey on Organic Farming 2007: Main Results. In: IFOAM/FiBL 2007: The World of Organic Agriculture. Bonn/Frick 2007. Available at http://orgprints.org/10506 /

Information Sources: Overview

HELGA WILLER[1]

In this chapter, some of the major sources for information related to organic farming are presented.

BioFach: Catalogue and BioFach Newsletter

NürnbergMesse, which runs the BioFach fair, publishes the BioFach Newsletter, which is a good source of information on organic farming developments worldwide. Furthermore, it publishes an annual catalogue of all exhibitors, which is a useful tool for market actors. The addresses are also available in a database, which can be accessed via the BioFach homepage.

- BioFach - World Organic Trade Fair, Nürnberg Messe, Messezentrum Nürnberg, 90471 Nürnberg, Germany, Internet: www.biofach.de
- BioFach-Newsletter, Internet: www.biofach.de/en/newsletter/archive/

Central Market and Price Reporting Bureau in Germany ZMP

The monthly information Bulletin of the German ZMP 'Oekomarkt Forum' has a news service that provides information about international developments in organic agriculture (in German). ZMP has also developed a database with data on European organic farming, funded by the European Commission (EISfOM project) and the German Federal Organic Farming Scheme.

- General information about the activities of ZMP related to organic farming is available at the ZMP homepage at www.zmp.de/agrarmarkt/branchen/oekomarkt.asp
- The ZMP organic market database is available at www.zmp.de/oekomarkt/Marktdatenbank/en/index.asp.

EkoConnect - International Center for Organic Agriculture of Central and Eastern Europe

EkoConnect is the International Center for Organic Agriculture of Central and Eastern Europe, and it is a member of the IFOAM. EkoConnect supports the exchange of information, knowledge and experiences in the field of organic agriculture. The organization also offers its network to people and their organizations engaged in the organic sector from Western and Eastern Europe. Since March 2005, EkoConnect has published a bi-monthly newsletter about organic agriculture in Central and Eastern Europe (in several languages).

- Homepage of the International Centre for Organic Agriculture of Central and Eastern Europe EkoConnect: www.ekoconnect.org/en_index.html

---

[1] Dr. Helga Willer, Communication, Research Institute of Organic Agriculture (FiBL), Ackerstrasse, 5070 Frick, Internet www.fibl.org

Eurostat Statistics on Organic Farming

Eurostat, the statistical office of the European Union, provides several data sets on organic agriculture for the EU countries. Eurostat also provides in depth reports about the statistical development of organic farming in Europe.

- Eurostat Homepage: http://epp.eurostat.ec.europa.eu/
- A link to the various Eurostat databases and documents related to organic farming is available via the Organic Europe homepage: www.organic-europe.net/europe_eu/statistics-eurostat.asp.

Food and Agriculture Organization of the United Nations (FAO)

The FAO offers a wide range of information on organic agriculture on its organic farming internet site. In 2005, the FAO set up the Organic Agriculture Information Management System (Organic-AIMS), with country information on organic agriculture. The site covers legal and institutional frameworks, institutions and experts. Selected documents are available.

The FAO statistical database FAOSTAT, which can be found at the FAO internet page, provides useful statistical information on agriculture worldwide.

- Food and Agriculture Organization of the United Nations (FAO), Organic Agriculture, Viale delle Terme di Caracalla, 00100 Roma, Italy. Internet: www.fao.org/organicag
- The general statistics are available at http://faostat.fao.org/default.aspx

Foreign Agricultural Service (FAS) of the United States Department of Agriculture (USDA): Reports & database

The attaché reports of the staff of the Foreign Agricultural Service (FAS) of the United States Department of Agriculture (USDA) provide in-depth information about organic farming in many countries of the world. The reports are available at the FAS Homepage.

- The attaché reports are available at www.fas.usda.gov/agx/organics/ attache.htm and www.fas.usda.gov/agx/organics/international.htm.
- The attaché reports database offers search options by products, countries and by date at www.fas.usda.gov/scriptsw/attacherep/default.asp.

International Federation of Organic Agriculture Movements (IFOAM)

The International Federation of Organic Agriculture Movements (IFOAM) is the international umbrella organization of organic farming organizations worldwide. It has about 700 members in more than 100 countries, which are listed in its membership directory (IFOAM 2007). The IFOAM homepage www.ifoam.org provides useful information and publications about organic farming worldwide. I

- International Federation of Organic Agriculture Movements (IFOAM), Charles-de-Gaulle-Str. 5, 53113 Bonn, Germany, Internet: www.ifoam.org

International Fund for Agricultural Development (IFAD)

The International Fund for Agricultural Development (IFAD) is a specialized agency of the United Nations, dedicated to eradicating rural poverty in developing countries. Its homepage is a useful tool to learn more about the agricultural situation in many developing countries. IFAD conducted two thematic evaluations of organic agriculture and poverty reduction: one covering Latin America and the Caribbean (IFAD 2002), and one covering Asia, primarily China and India (IFAD 2005).

- For more information about IFAD's work related to organic farming see
  www.ifad.org/evaluation/public_html/eksyst/doc/thematic/organic/organic.htm

International Trade Centre (ITC)

The organic farming homepage of the International Trade Centre (ITC) provides a wide range of information related to trade with organic products. Detailed information on the ITC activities is available in the following chapter by Alexander Kasterine.

- www.intracen.org/organics

Organic Market Info

This website offers a wide range of information on organic farming in Europe and worldwide with a special focus on market news.

- www.organic-market.info

Organic Monitor

Organic Monitor is a business research & consulting company that specializes in the global organic and related product industries. It provides a range of business services to organizations that are active in these industries. Services include business research publications, customized research and business consulting. The Organic Monitor homepage provides summaries of the research reports as well as market news.

- www.organicmonitor.com

The Organic Standard (TOS)

'The Organic Standard,' a magazine devoted to international certification issues, was launched in 2000. Published by Grolink, 'The Organic Standard' provides regular and up-to-date information on issues regarding organic farming worldwide, and a trial issue can be ordered via the internet site of the magazine.

- The Organic Standard www.organicstandard.com

Research Institute of Organic Agriculture (FiBL)

The FiBL Homepage www.fibl.org provides a wide range of information on the international work of the institution. Furthermore, FiBL maintains homepages of relevance to international organic farming including the Organic Europe homepage with country reports and addresses and the Organic World Homepage (links, currently being set up). FiBL also provides market studies.

- www.fibl.org, www.organic-europe.net and www.organic-world.net

References and further reading

Censkowsky, Udo, Ulrich Helberg, Anja Nowack, Mildred Steidle (2006) Overview of production and marketing of organic wild products. International Trade Centre (ITC), Geneva

El-Hage Scialabba, Nadia and Caroline, Hattam (Eds) (2002): Organic agriculture, environment and food security. Environment and Natural Resources Series 4. Food and Agriculture Organization of the United Nations FAO, Rome, Italy. Available at: www.fao.org/DOCREP/005/Y4137E/Y4137E00.htm

International Federation of Organic Agriculture Movements (IFOAM) (2007): Organic Agriculture Worldwide. IFOAM Directory of Member Organizations and Associates 2007. IFOAM, Bonn, Germany. Info available at http://shop.ifoam.org/bookstore/product_info.php?cPath=27&products_id=84

International Fund for Agricultural Development (IFAD) Evaluation Committee (2002): Thematic Evaluation of Organic Agriculture in Latin America and in the Caribbean. International Fund for Agricultural Development (IFAD) Evaluation Committee - Thirty-Second Session, Rome, 9 December 2002. Document available at www.ifad.org/gbdocs/eb/ec/e/32/EC-2002-32-W-P-3.pdf

International Fund for Agricultural Development (IFAD) / Fondo Internacional de Desarrollo Agrícola (FIDA), Unidad Regional de Asistencia Técnica (RUTA). Centro Agronómico Tropical de Investigación y Enseñanza (CA-TIE), Organización de las Naciones Unidas para la Agricultura y la Alimentación (FAO) (2003): Memoria del Taller: Agricultura Orgánica: Una herramienta para el desarrollo rural sostenible y la reducción de la pobreza (Organic Agriculture, a tool to sustainable development and poverty reduction). Del 19 al 21 de mayo de 2003, Turrialba,Costa Rica 2003. Available at www.fao.org/es/esc/common/ecg/30476_es_RUTAtaller.pdf

International Fund for Agricultural Development (IFAD) (2005) Organic Agriculture and Poverty Reduction in Asia: China and India Focus. Thematic Evaluation = Document of the International Fund for Agricultural Development. Rome. Report No. 1664. Info available at www.ifad.org/evaluation/public_html/eksyst/doc/thematic/organic/index.htm. Full document available at www.ifad.org/evaluation/public_html/eksyst/doc/thematic/organic/asia.pdf

International Trade Centre (ITC) (1999): Organic food and beverages: world supply and major European markets. ITC, CH-Geneva, info at www.intracen.org/mds/sectors/organic/welcome.htm

International Trade Centre, Technical Centre for Agricultural and Rural Cooperation and Food and Agriculture Organization of the United Nations (Eds.) (2001): World Markets for Organic Fruit and Vegetables - Opportunities for Developing Countries in the Production and Export of Organic Horticultural Products. Food and Agriculture Organization FAO, Rome. Available at www.fao.org/docrep/004/y1669e/y1669e00.htm

Parrot Nicholas and Terry Marsden (2002) The Real Green Revolution. Organic and agroecological farming in the South. Greenpeace, London, United Kingdom. Download at www. green-peace.org/multimedia/download/1/36088/0/realgreenrev.pdf

United Nations Conference on Trade and Development (2004): Trading Opportunities for Organic Food Products from Developing Countries. Strengthening Research and Policy Making Capacity on Trade and Environment in Developing Countries. United Nations, New York and Geneva. Available at http://r0.unctad.org/trade_env/test1/publications/organic.pdf
The report offers a general overview on trading opportunities for organic products from developing countries.

The Organics Trade Development Programme (OTDP) of the International Trade Centre (ITC)

ALEXANDER KASTERINE[1]

The International Trade Centre (ITC) is the joint technical cooperation agency of the United Nations Conference on Trade and Development (UNCTAD) and the World Trade Organization (WTO). It works with the organic sector in developing countries through its Organics Trade Development Programme (OTDP).

Developing countries face serious obstacles to the export of organic products, including meeting the quality demands of buyers, a lack of information about standards, dealing with the complexities and costs of certification, responding to market trends and establishing partnerships with potential importers. ITC works with small companies and trade support institutions in overcoming these obstacles and to increase access international markets. Through its Organics Trade Development Programme, ITC supports the development of the organic sector in developing countries through a number of channels, including the provision of market information, facilitating business-to-business contact, use of the internet, and developing trade through training in standards and certification and through policy support. The program is currently working in Kenya, Madagascar, Rwanda and Uganda. All UN Member States are eligible to join the program subject to availability of funding.

Export promotion

ITC runs a training program for organic conversion and marketing in the east and southern Africa, with the objective of helping Small and Medium Enterprises (SMEs) and farm cooperatives access national and international organic markets. During 2007, ITC held a number of training events, including a national seminar in conjunction with the Ministry of Agriculture of Rwanda. In April, ITC accompanied a group of eight European importers to Uganda to meet exporters and farmers, resulting in new deals for exporters. At BioFach 2008, ITC supported the participation of 20 companies from eastern and southern Africa to market their products at the African Pavilion, and another 10 from south Asia. These 30 exporters are supplied by thousands of farmers, and so there are high hopes that good deals can be arranged at BioFach!

Climate change

Climate change is one of the defining policy challenges of our time, and how agriculture responds to this is of key interest, particularly in developing countries. In order to investigate this further, ITC commissioned the Swiss Research Institute of Organic Agriculture (FiBL) to review the scientific evidence for organic agriculture's role in mitigation of and adaptation to climate change.

---

[1] Dr. Alexander Kasterine, Senior Market Development Adviser, International Trade Centre (ITC), UNCTAD/WTO, Rue de Montbrillant, 54-56, 1202 Geneva 10, Switzerland, Internet www.intracen.org/organics

Greenhouse gas emissions associated with organic products became a contentious issue in 2007 when the UK's Soil Association announced that it was considering a ban on certification for organic products airfreighted to the UK. This proposal was made on the grounds of high greenhouse gas emissions associated with airfreighted products. However, the ITC was concerned that the move posed a threat to the livelihoods of fruit and vegetable exporters in eastern and southern Africa, people with 'carbon footprints' 50 times smaller than an average European. The ITC, in partnership with UNCTAD, UNEP and DFID, initiated a joint campaign against the ban, which led to the Soil Association to accept certification of airfreighted products, albeit with a requirement for products from fair and 'ethical' sources; this decision safeguards the livelihoods of approximately 22'000 people involved in the organics export trade in Africa - for the time being.

Market information

The Organic Market News Service is a bimonthly publication for SMEs and trade support institutions in sub Saharan Africa. It carries information on prices, market trends and in-depth features on selected organic products and geographical focus areas. This is a new initiative from ITC launched in 2007.

In 2007/2008, ITC published the following studies, all of which are available as free downloads from www.intracen.org/organics:

- Organic Agriculture and Climate Change (Working Paper);
- World Market for Organic Wild Collection Products;
- The Economic Impact of a Ban on Imports of Airfreighted Products to the UK;
- Airfreight Transport of Fresh Fruit and Vegetables: A Review of the Environmental Impact and Policy Options; and
- The European Market for Organic Fruits and Vegetables from Thailand.

ITC's web portal Organic Link contains freely available database of importers and exporters of organic products, NGOs and research centers (www.intracen.org/organics), which enables buyers to find suppliers according to product and country of origin, and similarly, enables exporters to find new buyers. The website also serves as valuable source of freely available sector information, with links to several hundred market studies, business news and directories. A FAQ section gives users answers to frequently asked questions on organic and natural products.

The Global Market for Organic Food & Drink[1]

AMARJIT SAHOTA[2]

Introduction

Global demand for organic products remains robust with sales increasing by over 5 billion US Dollars[3] a year. Organic Monitor estimates international sales to have reached 38.6 billion US Dollars in 2006; this is over double that of 2000 when sales were at 18 billion US Dollars. Consumer demand for organic products is concentrated in North America and Europe; these two regions comprise 97% of global revenues. Other regions like Asia, Latin America and Australasia are important producers and exporters of organic foods.

The global organic food industry has been experiencing acute supply shortages since 2005. Exceptionally high growth rates have led supply to tighten in almost every sector of the organic food industry: fruits, vegetables, beverages, cereals, grains, seeds, herbs, spices, etc. The organic meat & dairy sector have been adversely affected because of the lengthy conversion process to organic methods and shortages of organic inputs. Organic feed shortages are preventing farmers to raise production levels. Indeed, a large rise in organic milk production in the US in spring 2007 was directly because of farmers completing their conversion period before a new ruling on organic feeds went into effect.

With supply lagging demand in most sectors, growth in the global organic food industry is expected to be stifled by supply shortages for a number of years. Organic Monitor expects supply-demand imbalances to continue throughout this decade.

Europe

Europe has the largest and most sophisticated market for organic food & drink in the world, valued at about 20 billion US Dollars in 2006. Its leadership is partly because of the depreciation of the US dollar in the foreign exchange; North America generated over half of global revenues until 2005. The growing weakness of the US dollar is expected to make Europe more prominent in the coming years. Even within Europe, organic food sales are concentrated with the bulk of sales coming from Western Europe. Indeed, four countries - Germany, France, Italy and the UK - comprise over 75 percent of regional revenues. Other countries like Denmark, Sweden and the Netherlands are showing high growth however they have much smaller markets because of their small consumer markets.

The German and UK markets continue to show the fastest growth in Europe. High growth

---

[1] This chapter has been prepared by updating information in The Global Market for Organic Food & Drink: Business Opportunities & Future Outlook (Organic Monitor, 2006). No part of this chapter maybe reproduced or used in other commercial publications without written consent from Organic Monitor. To request permission, write to: Organic Monitor, 20B The Mall, London W5 2PJ, Tel. +44 20 8567 0788, E-mail: postmaster@organicmonitor.com.
[2] Amarjit Sahota, Organic Monitor, 20B The Mall, London W5 2PJ, UK. Amarjit Sahota is the director of Organic Monitor, a business research & consulting company that specialises on the international organic & related product industries. More details are on www.organicmonitor.com
[3] 1 US Dollar = 0.79703 Euros. Average exchange rate 2006.

in the German market is partly because of the entry of the major retailers. Organic foods are becoming widely available in supermarkets, discount stores and drugstores. High demand is also continuing from organic food supermarkets and specialist retailers. Supply shortages in both the German and UK markets are stunting market growth rates, with retailers and food processors unable to find adequate supply of organic products.

Growth of the global market for organic food and drink 2002-2006 (Billion US Dollars)

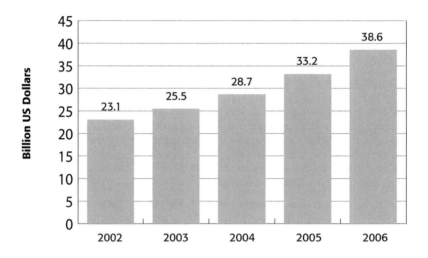

Source: The Global Market for Organic Food & Drink, Organic Monitor 2007

Figure 14: Growth of the global market for organic food and drink 2002-2006

Note: All figures are rounded. Source: Organic Monitor 2007

Scandinavian and Alpine consumers are the largest spenders on organic foods. The Swiss are the largest organic food consumers, with average spend of over 140 US Dollars per capita. The Danes, Swedes and Austrians are the next largest spenders. In contrast, Southern, Central & Eastern European consumers are the lowest spenders on organic foods.

There is a small but growing market for organic foods in new EU accession countries. Central & Eastern Europe (CEE) comprise about 2 percent of European revenues. Although organic food production is increasing in these countries, mostly organic primary crops are grown with most exported to Western Europe. Indeed, countries like Romania and Ukraine are becoming important sources of organic cereals & grains. The lack of organic food processing in CEE countries causes most finished goods to come into the region from the West.

North America

Organic food & drink sales in North America continue to surge, with retail sales estimated at 17.3 billion US Dollars in 2006. The US has the largest market for organic products in the world, worth over 16 billion US Dollars. Significant increases in organic farmland are making it a leading producer. Almost all types of organic crops are grown in the country, however, imports are necessary because of supply falling short of demand. Latin American countries like Mexico, Argentina and Brazil export large quantities of organic foods to the US market.

The introduction of Canadian national regulations for organic agriculture and organic foods is expected to give the market a boost. The US market has been reporting high growth since the USDA implemented the National Organic Program (NOP) in 2002. A major driver of market growth in both countries is increasing distribution in conventional grocery channels; large retailers like Wal-Mart, Safeway and Loblaws are expanding their organic product ranges with many introducing private label products.

The North American market continues to consolidate with a number of important mergers and acquisitions in 2007. Whole Foods Market has emerged as 'the supernatural' since it acquired its close rival Wild Oats in the spring. The leading natural & organic food companies in USA and Canada - Hain Celestial and SunOpta - continued their purchasing sprees by buying smaller firms. Both companies are expanding in Europe with similar acquisitions.

Competitive stakes are rising in North America. High market growth rates have attracted multinational food companies as well as conventional food retailers. Food companies like General Mills and Dean Foods have purchased dedicated organic food companies while others like Campbell Soup and Dole have launched organic lines. An important development is the opening of dedicated organic food retailers by large retailers. SuperValu is operating Sunflower Market stores and Publix Super Markets is expanding its GreenWise chain.

Asia

The Asian market continues to show high growth in terms of organic food production and sales. Organic crops are grown across the continent, with some countries becoming international suppliers of organic commodities. China is well-established as a global source of organic seeds, beans, herbs and ingredients. India, Thailand and the Philippines are also becoming important producers and exporters.

Retail sales were about 780 million US Dollars in 2006. Demand is concentrated in Japan, South Korea, Singapore, Taiwan and Hong Kong - the most affluent countries in the region. As in other parts of the world, demand is surpassing supply with large volumes of organic foods imported into each country.

The Asian market is showing high growth because of widening availability and rising consumer awareness. A growing number of conventional food retailers, especially those in the big cities, are introducing organic products. The number of dedicated organic food shops is also rising, with many new store openings in countries like Singapore, Malaysia and Taiwan. Some large food companies are also coming into the market and introducing organic lines.

Consumer awareness of organic foods is rising partly because of the high incidence of health scares in recent years. The scares, some involving foods, are raising consumer awareness of health issues and stimulating consumer demand for organic products. Important health scares were Avian flu and Severe Acute Respiratory Syndrome (SARS) and those involving foods included cola drinks (India) and tofu (Indonesia).

Although organic food sales are rising, consumer demand remains subdued partly because of the low spending power of most Asian consumers. Organic food prices are exceptionally high in some Asian countries. In Japan, Taiwan and Singapore, some organic foods are priced 4-5 times as much as non-organic foods. Since most finished organic products come in from countries like Australia and the US, distribution costs and import tariffs inflate product prices.

Oceania

Australasia is an important producer of organic foods, though it is not yet a large consumer. Valued at about 340 million US Dollars, the Australasian market for organic food & drink comprises less than one percent of global sales. Small consumer markets and export-focus of producers are responsible for the small market size.

Australia and New Zealand are leading exporters of organic products. Important exports include organic beef, lamb, wool, kiwi fruit, wine, apples and pears. Although exports continue to increase, the portion of exports to total production is in decline as internal markets for organic food & drink develop.

The Australasian market for organic products is growing at a steady rate. Most sales are of organic fresh products like fruit, vegetables, milk and beef, however, there is growing organic food processing. The number of mainstream food retailers selling organic products is increasing, while new organic food shops continue to open.

Other Regions

Production and sales of organic products is also increasing in other regions like Latin America and Africa. Organic crops are grown across the South American sub-continent. Countries like Argentina, Brazil and Chile have become important producers, however, over 90 percent of their organic crops are destined for export markets. Most organic food sales in these countries are from major cities like Buenos Aires and São Paulo. Organic food production in Africa is almost entirely for the export market. Middle-Eastern cities like Dubai, Riyadh and Kuwait City are becoming important consumers of organic products.

Conclusions

Global sales of organic food & drink surpassed the 40 billion US Dollar mark for the first time in 2007. With demand for organic foods outpacing supply, high growth rates are envisioned to continue in the coming years.

Most demand is concentrated in Europe and North America where organic food production is increasing at a relatively low rate. Farmers are showing low interest in organic farming because of rising prices of agricultural products, partly because of high fuel costs and growing interest in bio-fuel crops.

Distribution of global revenues by region 2006

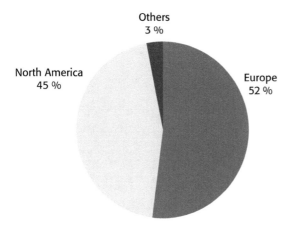

Source: Organic Monitor 2008

Figure 15: Distribution of global revenues by region 2006

Note: All figures are rounded. Source: Organic Monitor 2008

Organic agriculture in developing countries is rising at a much faster rate than that in the developed world. For instance, the amount of organic farmland has increased in triple digits in Africa, Asia and Latin America since 2000 whereas double-digit growth has been mostly observed in other regions. This disparity between production and consumption is putting the global organic food industry in a fragile condition. A dip in demand from Europe and / or North America would have a major impact on global production of organic foods. The industry could lose confidence as export markets close, resulting in oversupply and a decline in organic food prices.

Organic food producers in regions like Asia, Africa and Latin America are advised to become less reliant on exports and develop internal markets for their products. By developing local markets, producers can spread the business risk of organic food production. Consumers can also benefit by having access to regionally produced organic foods. The conversion rate towards organic farming in North America and Europe also needs to keep pace with organic food sales. Only when production and demand is more evenly spread can the organic food industry be considered truly global.

Reference

Organic Monitor (2006): The Global Market for Organic Food and Drink: Business Opportunities & Future Outlook. Organic Monitor, London

57

Standards and Regulations

BEATE HUBER[1], OTTO SCHMID[2], LUKAS KILCHER[3]

Introduction

Farmers' associations developed the first standards for organic production in the middle of the last century. The first international standards were published by IFOAM in 1980. The first regulations were passed by some European countries, including Austria and France, in the 1980s. In 1991, the EU passed the organic Regulation 2092/91 and set standards with major implications for international trade; they covered not only production standards, but standards for labeling and inspection as well. Various countries in Europe, Latin America and Asia - including Japan - introduced legislation in the 1990s. In 1999, the Codex Alimentarius approved the first guidelines for organic plant production, which were amended to include livestock production in 2001. In the new millennium, most major economies have established a regulation for organic production, including the Indian National Programme for Organic Production (NPOP), which passed in 2001, the US National Organic Program (NOP) that came into force in 2002, the Chinese legal framework, which was finalized in 2005 and the Canadian legislation that passed at the end of 2006.

In 2007, the EU passed the first part of a completely revised Regulation on Organic Production that will come into force on January 1, 2009. The implementation rules referring to the revised regulation are expected to be published in mid 2008. Many European States that are not members of the EU will start adapting their regulations to the EU Regulation in 2008/2009.

A full list of countries with regulations or in the process of drafting regulations on organic agriculture is below.

Table 7: Countries with regulations on organic agriculture

Region	Country	Remark	Website (where available)
European Union (27)	Austria	Fully implemented	eur-lex.europa.eu/LexUriServ/site/en/ consleg/1991/R/01991R2092-20070101-en.pdf
	Belgium	Fully implemented	As above
	Bulgaria	Fully implemented	As above
	Cyprus	Fully implemented	As above
	Czech Republic	Fully implemented	As above
	Denmark	Fully implemented	As above
	Estonia	Fully implemented	As above
	Finland	Fully implemented	As above
	France	Fully implemented	As above
	Germany	Fully implemented	As above
	Greece	Fully implemented	As above
	Hungary	Fully implemented	As above

[1] Beate, Huber, Research Institute of Organic Agriculture (FiBL), Ackerstrasse, CH-5070 Frick, Internet www.fibl.org
[2] Otto Schmid, Research Institute of Organic Agriculture (FiBL), Ackerstrasse, CH-5070 Frick, Internet www.fibl.org
[3] Lukas Kilcher, Research Institute of Organic Agriculture (FiBL) Ackerstrasse, CH-5070 Frick, Internet www.fibl.org

Region	Country	Remark	Website (where available)
	Ireland	Fully implemented	As above
	Italy	Fully implemented	As above
	Latvia	Fully implemented	As above
	Lithuania	Fully implemented	As above
	Luxembourg	Fully implemented	As above
	Malta	Fully implemented	As above
	Poland	Fully implemented	As above
	Portugal	Fully implemented	As above
	Romania	Fully implemented	As above
	Slovak Republic	Fully implemented	As above
	Slovenia	Fully implemented	As above
	Spain	Fully implemented	As above
	Sweden	Fully implemented	As above
	The Netherlands	Fully implemented	As above
	United Kingdom	Fully implemented	As above
Others Europe (11)	Albania	Not fully implemented	
	Croatia	Fully implemented	
	Iceland	Fully implemented	www.landbunadarraduneyti.is/log-og-reglugerdir/Reglugerdir/Allar_reglugerdir/nr/79
	Kosovo	Not fully implemented	
	Macedonia	Fully implemented	
	Moldova	Fully implemented	
	Montenegro	Fully implemented	www.skupstina.cg.yu/skupstinaweb/tekstovi_list.php?s_id_zakoda=110
	Norway	Fully implemented	
	Serbia	Not fully implemented	
	Switzerland	Fully implemented	www.admin.ch/ch/d/sr/c910_18.html
	Turkey	Fully implemented	?
Asia and Pacific Region (11)	Australia	Only export regulations	www.affa.gov.au/corporate_docs/publications/word/quarantine/approg/nationalstandard2.doc.
	Bhutan	Not fully implemented **)	
	China	Fully implemented	
	Georgia	Not fully implemented	
	India	Only export regulations *)	National Programme for Organic Production (NPOP) www.apeda.com/organic/index.html
	Israel	Only export regulations	
	Japan	Fully implemented	JAS Standards for organic plants and organic processed foods www.maff.go.jp/soshiki/syokuhin/hinshitu/e_label/specificJAS-organic.htm
	New Zealand	Only export regulations	New Zealand Food Safety Authority (NZFSA) Official Assurance Programme for Organic Products www.nzfsa.govt.nz/organics/index.htm
	Philippines	Not fully implemented	
	Korea South	Fully implemented	
	Taiwan	Fully implemented	
	Thailand	Fully implemented	Homepage of the National Bureau of Agricultural Commodity and Food Standards www.acfs.go.th/eng/index.php
The Americas & Caribbean (17)	Argentina	Fully implemented	
	Bolivia	Not fully implemented	www.aopeb.org/
	Brazil	Fully implemented	www.planetaorganico.com.br
	Canada	Not fully implemented	

Region	Country	Remark	Website (where available)
	Chile	Fully implemented	
	Costa Rica	Fully implemented	www.mag.go.cr/doc_d/reg_ley_mag.html
	Colombia	Fully implemented	
	Dominican Republic	Not fully implemented	
	Ecuador	Fully implemented	www.sica.gov.ec/agronegocios/productos%20para%20invertir/organicos/principal.htm
	El Salvador	Not fully implemented	www.elsalvadororganico.com.sv/
	Guatemala	Fully implemented	
	Honduras	Not fully implemented	www.senasa.gob.hn
	Mexico	Not fully implemented	
	Paraguay	Not fully implemented	
	Peru	Not fully implemented	
	Uruguay	Not fully implemented	
	USA	Fully implemented	www.ams.usda.gov/nop/indexIE.htm
Africa (3)	Ethiopia	Not fully implemented	
	Ghana	Not fully implemented	
	Tunisia	Fully implemented	

*) National regulation expected to be implemented in 2008.
**) National framework, but no labeling regulation.
Source: Huber, Silva, Gelman, FiBL 2006, updated 2008

Table 8: Countries in the process of drafting regulations

Region	Country	
Europe (3)	Bosnia & Herzegovina	
	Russia	
	Ukraine	
Asia and Pacific Region (8)	Armenia	
	Azerbaijan	
	Hong Kong	
	Indonesia	
	Jordan	
	Lebanon	
	Saudi Arabia	
	Vietnam	
The Americas & Caribbean (3)	Cuba	
	Nicaragua	
	St. Lucia	
Africa (7)	Cameroon	
	Egypt	
	Kenya	
	Madagascar	
	South Africa	www.afrisco.net/Html/Product_Stardards.htm
	Tanzania	
	Zambia	

Source: Huber, Silva, Gelman, FiBL 2006 updated 2008

The data on regulations around the world have been collected from authorities and experts. The classification of whether the regulation is 'not yet fully' or 'fully implemented' is based feedback of the persons interviewed, but was not subject to verification. We received responses from experts and authorities in 60 percent of the countries. It may be assumed that a majority of the 40 percent of non-responding countries did not pass legislation on organic production, although the share of countries in the process of developing legislation is

61

probably greater than reflected. Please send comments or information on countries not listed to beate.huber@fibl.org.

International standards & regulations

IFOAM Basic Standards

The IFOAM Basic Standards[1] define how organic products are grown, produced, processed and handled. They reflect the current state of organic production and processing methods. The IFOAM Basic Standards - together with the IFOAM Accreditation Criteria - constitute the IFOAM Norms, which provide a framework for certification bodies and standard-setting organizations worldwide to develop their own certification standards; the IFOAM Norms are often referred to as 'standards for standards.' In close cooperation and consultation with IFOAM member organizations and other stakeholders, the IFOAM Standards Committee develops the IBS. The IFOAM Basic Standards are presented as general principles, recommendations, basic standards and derogations. In 2005, the General Assembly of IFOAM adopted the four overarching Principles of Organic Agriculture - the principles of health, ecology, fairness and care. The IFOAM Organic Guarantee System is currently under revision, including a complete revision of the IFOAM Norms. In 2006 and 2007, the IFOAM World Board and Standards Committee organized two rounds of consultation on the draft version of the IFOAM Standards. These draft standards will be put to a vote after discussion at the General Assembly in June 2008 in Modena, Italy.

The Codex Alimentarius Guidelines

The need for clear and harmonized rules has not only been taken up by private bodies, IFOAM and state authorities, but also by United Nations Organizations, including the Food and Agriculture Organization of the United Nations (FAO) the World Health Organization (WHO) and the United Nations Conference on Trade and Development (UNCTAD). The FAO and WHO consider international guidelines on organically produced food products to be important instruments for consumer protection and to facilitate trade. They also provide assistance to governments wishing to develop regulations in this area, in particular in developing countries and countries in transition.

The Codex Alimentarius Commission was created in 1963 by the FAO and WHO to develop food standards, guidelines and related texts such as codes of practice under the Joint FAO/WHO Food Standards Program; it established the Guidelines for the Production, Processing, Labeling and Marketing of Organically Produced Foods. The Codex Commission approved plant production guidelines in June 1999, and animal production guidelines in

---

[1] On the homepage of IFOAM http://www.ifoam.org under 'Organic Guarantee System,' the IFOAM Norms, consisting of the IFOAM Basic Standards for Organic Production and Processing and the IFOAM Accreditation Criteria for Bodies certifying Organic Production and Processing may be purchased. The website also provides information on the IFOAM Accreditation Program (see next chapter).

July 2001.[1] The requirements in these Codex Guidelines are in line with the IFOAM Basic Standards and the EU Regulation 2092/91. There are, however, differences with regard to the details and some specific areas that are covered by the varying standards.

The trade guidelines on organic food take into account the current regulations in several countries, in particular the EU Regulation 2092/91, private standards applied by producer organizations - especially those based on the IFOAM Basic Standards. These guidelines define the nature of organic food production and prevent claims that could mislead consumers about the quality of the product or the way it was produced.

From IFOAM's perspective, the Codex Guidelines are an important step towards the harmonization of international rules that serve to build consumer trust. They will be important for equivalence judgments under the rules of WTO. In terms of developing the market for organically produced food, these Codex Guidelines provide guidance to governments in developing national regulations for organic food.

The annex lists, which define what substances can be used in organic systems, have been under revision since 2005, with a focus on food processing and amended criteria for the use of new substances. A working group within the Codex Committee for Food Labeling (CCFL), which meets every year in May and is supported by the government of Canada, is charged with this work. The Codex Commission adopted the amendments the annex lists that were proposed by the CCFL in July 2007. In 2008, the discussion about a few controversial additives for food processing will continue.

EU Regulation 2092/91

Revision of the basic rules

In their conclusions on the European Action Plan on Organic Agriculture in 2004, the European Council requested that the EU Regulation 2092/91 on organic farming be revised to achieve a simplified, more coherent and less detailed regulation. In July 2007, 'Council Regulation (EC) No 834/2007 of 28 June 2007 on organic production and labelling of organic products and repealing Regulation (EEC) No 2092/91' was adopted. This regulation describes the objectives, principles and basic requirements of regulations for organic production. It does not cover most of those sections that are currently regulated by annexes I to VIII, including production rules, minimum inspection rules, list of approved farm inputs, ingredients, aids and additives. Detailed proposals for the revision of the technical annexes are expected to be published at the beginning of 2008 and to be adopted by mid 2008.

The new regulation has been completely restructured and rephrased. It replaces EU Regulation 2092/91 and will come into force on January 1, 2009. Although the Commission aimed at maintaining the main content of the current regulation, there are plenty of changes in the details, some of which may have serious impacts on organic farming in future.

---

[1] Information about Codex Alimentarius is available via the homepage www. codexalimentarius.net. The Codex-Alimentarius-Guidelines on organic agriculture can be downloaded from www.codexalimentarius.net/download/standards/360/CXG_032e.pdf.

Table 9: Content of the new EU Regulation on organic production

EU Regulation No 834/2007				Remarks and Comparison with EU Regulation No 2092/91
Title	Article			
TITLE I AIM, SCOPE AND DEFINITIONS	Article 1 Aim and scope			The new regulation defines the aim of the regulation. The scope slightly differs from Council Regulation (EEC) No 2092/91, in that is specifies that products from the hunting and fishing of wild animals shall not be considered as organic, and that mass catering shall not be subject to the regulation. Aquaculture and yeast have been added within the scope of the regulation.
	Article 2 Definitions			Definitions of the terms used in the regulation.
TITLE II OBJECTIVES AND PRINCIPLES FOR ORGANIC PRODUCTION	Article 3 Objectives			Definition of the objectives of organic production (not covered in EU Regulation 2092/91).
	Article 4 Overall principles			Definition of the principles of organic production (not covered in EU Regulation 2092/91): (a) sustainable management system for agriculture (b) Restriction of the use of external inputs. (c) strict limitation of chemically synthesized inputs to exceptional cases. (d) Flexibility in the application of rules for regional differences in climate and local conditions and stages of development.
	Article 5 Specific principles applicable to farming			Not covered in EU Regulation 2092/91.
	Article 6 Specific principles applicable to processing of organic food			Not covered in EU Regulation 2092/91
	Article 7 Specific principles applicable to processing of organic food			Not covered in EU Regulation 2092/91.
TITLE III PRODUCTION RULES	**Chapter 1 General production rules**	Article 8 General requirements		The general requirements for organic production are now described in the main rule and not in the annexes as in EU Regulation 2092/91
		Article 9 Prohibition on the use of GMOs		GMOs and products deriving from GMOs are prohibited. There is no separate threshold for organic production. and as a result the general EU threshold of 0.9 % is applied.
		Article 10 Prohibition on the use of ionizing radiation		The use of ionizing radiation is prohibited.
	Chapter 2 Farm production	Article 11 General farm production rule		The entire agriculture holding shall be managed organically. Split production is possible under specific circumstances if different animal species (except fish) and plant varieties that are easy to distinguish are involved.
		Article 12 Plant production rules		No severe change to production rules of EU Regulation 2092/91. Prohibition of mineral nitrogen fertilizer, approval of biodynamic preparations, reference to sustainability and environment protection. Criteria for farm inputs are defined.
		Article 13 Production rules for seaweed		New area. Not covered under EU Regulation 2092/91.
		Article 14 Livestock production rules		No severe change to production rules of EU Regulation 2092/91.
		Article 15 Production rules for aquaculture animals		New area. Not covered in EU Regulation 2092/91.

EU Regulation No 834/2007			Remarks and Comparison with EU Regulation No 2092/91
Title	Article		
		Article 16 Products and substances used in farming and criteria for their authorization	Not covered in EU Regulation 2092/91.
		Article 17 Conversion	Definition of rules for conversion to organic farming. The conversion period shall be defined in the implementation rules.
	Chapter 3 Produc-tion of processed feed	Article 18 General rules on the production of processed feed	Slightly revised to current rules.
	Chapter 4 Production of processed food	Article 19 General rules on the production of processed food	Slightly revised to current rules.
		Article 20 General rules on the production of organic yeast	Not covered in EU Regulation 2092/91.
		Article 21 Criteria for certain products and substances in processing	Not covered in EU Regulation 2092/91. Clarification for calculation of salt and water. Dietary requirements are a justification for listing additives. Processing procedures shall be defined implementation rules. Definition of criteria for approval of additives and aids.
	Chapter 5 Flexibility	Article 22 Exceptional produc-tion rules	Exceptions to the regulation may be granted in the following cases, with specific details regulated in the implementation rules: Climatical, geographical or structural constraints; To ensure access to farm inputs; Lack of organic ingredients; Management of animal husbandry; Catastrophic circumstances; Lack of non-GMO food and feed additives and aids. Not covered in EU Regulation 2092/91.
TITLE IV LABELLING		Article 23 Use of terms referring to organic production	Major changes compared to EU Regulation 2092/91: Minimum 95 % organic ingredients, no longer 70 % provision Labeling of organic ingredients possible for products containing conventional ingredients if only approved additives and aids are used. In this case, the share of conventional ingredients may be higher than 5 %..
		Article 24 Compulsory indications	Major changes compared to EU Regulation 2092/91: Obligatory use of the EU logo for products produced in EU; EU logo may be used for products produced outside of the EU; Indication of origin obligatory ('EU-' or 'non-EU-Agriculture') if EU logo is used.
		Article 25 Organic production logos	EU logo may be used in labeling, presentation and advertising. National or private logos may also be used as long as the provisions of the regulation are satisfied.
		Article 26 Specific labeling requirements	Labeling rules for organic feed, in-conversion products of plant origin and seeds shall be defined in the implementation rules.

EU Regulation No 834/2007		Remarks and Comparison with EU Regulation No 2092/91
Title	Article	
TITLE V CONTROLS	Article 27 Control system	System of controls implemented by the Member States shall be in conformity with Regulation 882/2004 on official controls for food and feed security.
	Article 28 Adherence to the control system	Operator must notify organic production with competent authority. Rules are similar to current regulation.
	Article 29 Documentary evidence	Each certified operator shall receive 'documentary evidence' on the certification. Details will be defined in the implementation rules. Advantage of electronic certification shall be taken into account.
	Article 30 Measures in case of infringements and irregularities	Similar to current rules.
	Article 31 Exchange of information	Competent authorities and control bodies must exchange information on results of controls where applicable.
TITLE VI TRADE WITH THIRD COUNTRIES	Article 32 Import of compliant products	Complete revision of import rules, see chapter below.
	Article 33 Import of products providing equivalent guarantees	Complete revision of import rules, see chapter below.
TITLE VII FINAL AND TRANSITIONAL RULES	Article 34 Free movement of organic products	Administrative Rules
	Article 35 Transmission of information to the Commission	
	Article 36 Statistical information	
	Article 37 Committee on organic production	
	Article 38 Implementing rules	
	Article 39 Repeal of Regulation (EEC) No 2092/91	
	Article 40 Transitional measures	
	Article 41 Report to the Council	
	Article 42 Entry into force and application	

Information on the revision of the EU regulation is available on the revision info page of the IFOAM EU Group.[1] An updated consolidated version of the EU Regulation 2092/91 and the revised Regulation 834/2007 are published on the EUR-Lex website. They are available in all official languages of the European Union[2].

---

[1] www.ifoam.org/about_ifoam/around_world/eu_group/web_Revision/Revision_info_page.html
[2] Current regulation: Council Regulation (EEC) No 2092/91 of 24 June 1991 on organic production of agricultural products and indications referring thereto on agricultural products and foodstuffs and amendments: http://eur-lex.europa.eu/LexUriServ/site/en/consleg/1991/R/01991R2092-20070101-en.pdf and Council Regulation (EC) No 834/2007 of 28 June 2007 on organic production and labelling of organic products and repealing Regulation (EEC) No 2092/91 http://eur-lex.europa.eu/LexUriServ/site/en/oj/2007/l_189/l_18920070720en00010023.pdf

The EU has announced a further revision of the current EU logo. The new design is expected to be published in 2008.

Revised import procedures

At the end of December 2006, the EU published new regulations concerning the importation of organic products. The revised import procedures replace the current (temporary) system of import authorizations by an approval system for inspection bodies operating in countries outside of the EU. The existing system for approval of countries on the so-called 'Third Country List' will be maintained. The implementation rules of the new regulation are expected to be passed in 2008. The implementation of the new import rules will presumably be delayed and only fully implemented in 2010.

To import products into the EU, the products must have been certified by an inspection body or authority recognized by the European Commission. The EU will publish lists of approved inspection bodies and authorities as well as approved third countries. There will be three different lists:

1) List of inspection bodies that have been accredited according to EN 45011/ISO 65 and that apply an inspection system and production rules compliant with the EU Regulation.
    The provision on compliance with the EU regulation is new. So far, the EU only requested equivalency with the EU requirements. It will have to be determined what compliance means with regard to other climate conditions such as the tropics or in relation to Internal Control Systems that are accepted by the EU in developing countries.
2) List of inspection bodies that apply an inspection system and production standards equivalent to the EU regulation.
    The previous EU legislation requested 'equivalency' with the production and inspection provisions of the EU regulation for imported products. It does not require accreditation of the certification bodies and allows certification bodies to apply their own standards as long as they are 'equivalent' with EU provisions.
3) List of countries whose system of production complies with rules equivalent to the EU production and inspection provisions.
    This list corresponds to the existing Third Country List, and procedures to become listed will presumably remain very similar.

Under option 1) and 2) the inspection bodies can either be located within or outside the EU. Inspection bodies or authorities shall provide to the EU the assessment reports issued by the accreditation body or, as appropriate, the competent authority on the regular on-the-spot evaluation, monitoring and multi-annual re-assessment of their activities. The EU provides the option to assign experts to conduct 'on-the-spot' examinations and shall ensure appropriate supervision of the recognized inspection bodies by regularly reviewing their recognition. It can be assumed that the EU will not itself supervise the inspection bodies, but will rely on the reports provided by competent bodies such as the national accreditation bodies or the International Organic Accreditation Service (IOAS).[1]

---

[1] For information on the International Organic Accreditation Service see their homepage at www.ioas.org. A list of the IFOAM accredited certification bodies is available at www.ioas.org/acbs.htm

Under options 2) and 3) (equivalency-option), the imported products have to be covered by a certificate of inspection, which is not a provision under option 1). For option 2) and 3), Codex Alimentarius shall be taken into account for assessing equivalency.

The new import regulation provides for a more consistent and effective control system for imported products, and improves the options for supervising inspection bodies operating in Third Countries. Furthermore, it increases transparency by publishing lists of recognized inspection bodies. In the current system, it was difficult for inspection bodies outside the EU to prove the acceptance of their certification in the EU. They depended on European importers willing to take the hurdle to apply for an import authorization with a new or unknown inspection body. The new system allows inspection bodies from non-EU-countries to apply for recognition at their own initiative; they can prove their recognition prior to the start of trade relationships. This reduces the risk of importers when importing products certified by non-European and/or less known inspection bodies.

Import requirements of major economies

The most important import markets for organic products are the EU, US and Japan. All of them have strict regimes for the importation of organic products. Recently, local markets are growing steadily in other regions, and as a result, products are imported in other regions. Some of these emerging markets have developed rules for importing organic products, although most of them are not yet (fully) implemented, including in China and Canada. In the EU, US and Japan, products may only be imported if the certifying agency has been approved by the respective competent authority. Approval of certification bodies requires compliance or equivalency with the requirements of the importing countries, which can be achieved by the following options:

- Bilateral agreements between the exporting and the target import country
- Direct acceptance of the certifying agency by the target import country

Bilateral agreements between the exporting and the target import country:

Most importing countries - including the US, EU and Japan - have options for bilateral recognition; a country may confirm that another country's control system and the standards are in line with the national requirements, and that the products certified in those countries can be sold on the national market. The US, however, has not signed any full bilateral recognition agreements. The EU currently recognizes seven countries[1]. The bilateral agreements are largely political agreements that depend on the will and political negotiations of the governments, rather than the results of technical assessments.

Thus far, the US has accepted few foreign governments' accreditation procedures. Certification bodies accredited according to the US requirements by Denmark, Great Britain, India, Israel, New Zealand and Quebec are accepted by the USDA for certifying according to the US National Organic Program NOP[2] without being directly accredited by United States De-

---

[1] Argentina, Australia, Costa Rica, New Zealand, India, Israel, Switzerland
[2] National Organic Programme (NOP) www.ams.usda.gov/nop/indexIE.htm

partment of Agriculture (USDA). This is just recognition of the accreditation procedures; the respective certification bodies still have to meet the requirements of NOP to issue certificates accepted by the US.

In addition, the US is negotiating equivalency agreements with Australia, the European Union, India and Japan. This means that USDA would determine that their technical requirements and conformity assessment system adequately fulfill the objectives of the NOP, and double certification would not be necessary for imports. The US announced that equivalency determinations are very complex and time consuming, and for the time being, negotiations with the EU are on hold, especially for animal production.

Acceptance of the certifying agency by the target import country

The US, EU and Japan have options for recognizing certification bodies operating outside the country. The technical requirements for achieving such recognition are difficult to meet, and the associated fees are high. Maintaining recognition and/or the necessary accreditation requires substantial financial capacity and personnel from the certification agency.

The US National Organic Program (NOP) requires all produce labeled as organic in the US to meet the US standards, including imported products. The US system provides for the approval of certification bodies as agents to operate a US certification program. Retroactive certification is not possible. Inspections have to be conducted by inspectors trained in NOP requirements using NOP questionnaires, and only certificates issued by certification bodies accredited by the US Department of Agriculture USDA are accepted. It is not relevant whether the certification body is based in the US or elsewhere. So far, almost 100 certification bodies have been accredited according to NOP by the USDA, and only produce certified by these certification bodies may be exported to the US.

Private Standards

In some European countries, farmers' associations had already formulated their private standards and labeling schemes long before national regulations came into force. These quality marks or logos, for example in the UK, in Denmark, Austria, Sweden and Switzerland, are well trusted by consumers and are one of the reasons for the current boom in the market for organic products in these countries.

Compared to national regulations, private standards are developed from the bottom up rather than imposed from above. However, since the implementation of national regulations, private standards are forced to comply, and state authorities increasingly make standards decisions instead of farmers' associations. There are still some areas in which private standards have additional requirements or cover specific areas more in detail, such as compared with the 834/2007 (Schmid et al. 2007, see also database www.organicrules.org). In some countries, these private standards have relevancy for a specific market.

In 2002, UNCTAD, the FAO and IFOAM initiated the International Task Force on Harmonization and Equivalence in Organic Agriculture (ITF) to harmonize organic standards and regulations. This partnership between the private organic community and the United Nations offers a forum for public and private discussions, and aims to initiate the development of a constructive and effective partnership between the private and the public sector. In 2007, the draft for International Requirements for Organic Certification Bodies (IROCB) was reviewed according to comments received by stakeholders, and the first draft of a Tool for Equivalence of Organic Standards and Technical Regulations (EquiTool) was discussed.

Relationship to Fair Trade

Many producer associations in the emerging markets and markets in transition conform to the requirements of the Fair Trade organizations such as the Fair Trade Labeling Organization International (FLO), Transfair, Max Havelaar and World Shops (Weltlaeden). Having a Fair Trade label does not necessarily mean, however, that the products can also be sold as organic. Fair Trade products must also be certified according to accredited organic inspection procedures to be labeled as organic. IFOAM maintains close contact with FLO and its members, due to the fact that many projects conform to the standards of both organizations. The combination of organic and Fair Trade labeling can enhance a product's marketability. Additional information and regulations can be downloaded at www.flo-international.org.

Literature

Commins, Ken (2003): Overview of current status of standards and conformity assessment systems, Discussion Paper on the International Task Force on Harmonization. http://orgprints.org/3109/

DiMatteo, Katherine (2007) Overview of group certification. Prepared for the 7th meeting of the UNCTAD/ FAO/ IFOAM International Task Force on Harmonisation and Equivalence in Organic Agriculture in Bali, Indonesia. 27-30 November 2007. http://www.unctad.org/trade_env/itf-organic/meetings/itf7/ITF0711_GrowerGroups.pdf

Huber et al. (2007) Standards and Regulations. In: Willer/Yussefi (Eds.) The World of Organic Agriculture. Statistics and Emerging Trends 2007, International Federation of Organic Agriculture Movements IFOAM, Bonn, Germany and Research Institute of Organic Agriculture (FiBL), Frick, Switzerland. orgprints.org/10506/

Kilcher Lukas et al. (2004): The Market for Organic Food and Beverages in Switzerland and the European Union. Overview and market access information, pp 156, Forschungsinstitut fuer biologischen Landbau (FiBL) und Swiss Import Promotion Program (SIPPO), Second Edition Frick/Zuerich

Schmid Otto et al. (2007): Analysis of EEC regulation 2092/91 in relation to other national and international organic standards. Report (Deliverable D 3.2). EEC 2092/91 (Organic) Revision project. Research Institute of Organic Agriculture (FiBL), CH-Frick. www.organic-revision.org/pub/D_3_2_final%20report_low.pdf. Database www.organicrules.org

Websites

www.fao.org/organicag/: Information on organic agriculture by FAO with detailed country reports including the on legal situation

www.ifoam.org/about_ifoam/standards/index.html: IFOAM Guarantee system

www.organic-europe.net/: Extensive country reports and address database

www.ioas.org: Homepage of the International Organic Accreditation Service

www.ams.usda.gov/nop/indexIE.htm: Information about the US National Organic Program (NOP)

organicrules.org/: Database which compares national and private standards with the EU Regulation 2092/91.

www.unctad.org/trade_env/itf-organic/ welcome1.asp: International Task Force on Harmonization and Equivalency in Organic Agriculture (ITF).

www.codexalimentarius.net/ download/standards/360/CXG_032e.pdf The Codex Alimentarius Commission and the FAO/WHO Food Standards Programme: Organically Produced Foods, Rome 2007

ec.europa.eu/agriculture/qual/organic/ : General EU Commission internet site for organic farming in all EU languages

www.ifoam.org/about_ifoam/around_world/eu_group/web_Revision/Revision_info_page.html : IFOAM EU Info page on the Revision process of EU Regulation 2092/91.

East African Organic Products Standard and more[1]

SOPHIA TWAROG[2]

In East Africa, inter-agency collaboration helped secure a major breakthrough in the adoption of the new East African Organic Products Standard (EAOPS) and energized vibrant regional public private sector partnership in this exciting growth sector.

In 2005, private-sector driven growth in production and exports of certified organic products in Kenya, Uganda and the United Republic of Tanzania was improving livelihoods of thousands of smallholder farmers. Yet only a small handful of government policymakers were aware of the benefits provided by organic agriculture. Organic agriculture was invisible in public policies, plans, budgets, research and extension services. Public-private sector dialogue and regional cooperation were the exception rather than the rule.

Lack of cooperation and harmonization was becoming a possible threat to further development of the sector. There were at least five public or private standards for organic agricultural production in East Africa. Stakeholders were concerned that this multitude of standards could eventually become a technical barrier to trade within the region and place unneeded restrictions on regional collaboration. There was general consensus that the time was ripe for the development of a common East African organic standard.

Under the framework of the UNEP[3]-UNCTAD[4] Capacity Building Task Force on Trade, Environment and Development (CBTF), national multistakeholder consultations were held in the three countries to assess needs and aspirations. A project was then developed which included research on key topics such as current state of the sector in East Africa, organic agriculture and food security, and best practices for organic policies-a practical toolkit for policymakers. The project financed national integrated assessments in the three countries and facilitated regional cooperation and exchange. The European Commission and the Swedish International Agency for Development Cooperation (Sida) provided financial support.

Other partners and countries were brought in - The International Federation of Organic Agriculture Movement (IFOAM) with its wealth of knowledge on organic agriculture, the International Trade Centre (UNCTAD/WTO), the Export Promotion of Organic Products from Africa (EPOPA) program, and new East African Community members Rwanda and Burundi. The 'East African Organic Team' founded at a CBTF workshop in March 2006 includes all public and private sector actors and international partners joined in the common cause of promoting the organic sector in East Africa. It is a vibrant regional and inter-

---

[1] This article is based largely upon information prepared by the author for the *UNCTAD Annual Report 2007* (UNCTAD/DOM/2007/3).

[2] Dr. Sophia Twarog, Economic Affairs Officer, Trade, Environment and Development Branch, UNCTAD/DITC, E.8015, Palais des Nations, 1211 Geneva 10, Switzerland, www.unctad.org/trade_env

[3] UNP is the United Nations Environment Programme, www.unep.org

[4] UNCTAD is the United Nations Conference on Trade and Development, www.unctad.org

national public-private sector partnership.

In all countries much momentum has been created - organic policies are being rapidly developed, national public-private sector dialogue is strong, and intensive exchanges are taking place at regional level.

A key achievement has been the development and adoption of the East African Organic Products Standard (EAOPS) through an intensive, inclusive and transparent regional consultation process. Feedback from national consultations, open calls for comments, field testing and technical comparisons with international organic standards were fed into the work of the Regional Standard Technical Working Group (RSTWG), the body tasked with writing the standard's text. RSTWG members comprise representatives of the National Standards Bodies, national organic movements and organic certifying bodies of Kenya, Tanzania, Uganda, Burundi and Rwanda, and the East African Business Council. The RSTWG completed its work in four meetings between October 2005 and December 2006. The EAOPS was adopted by the East African Council of Ministers in April 2007 and launched together with the associated East African Organic Mark (see picture) by the Prime Minister of Tanzania in May 2007.

The EAOPS is the second regional organic standard in the world after the European Union's and the first ever to have been developed in cooperation between the organic movements and the National Standards Bodies. The EAOPS is expected to boost organic trade and market development in the region, raise awareness about organic agriculture among farmers and consumers, and create a unified negotiating position that should help East African organic farmers win access to export markets and influence international organic standard setting processes.

Number of Organic Certifiers Jumps to 468

GUNNAR RUNDGREN[1]

In 2006, The Organic Certification Directory[2] contained 395 entries. The increase in numbers is mainly a result of a massive growth in South Korea (from 2 to 33 in one year) and Japan. In Japan, the number is almost back to the level for 2005, when there were 69 organizations. In most other countries, the situation is fairly stable.

Table 10: The countries with the largest number of certification bodies

	2007	2006	2005
USA	60	59	60
Japan	55	35	69
South Korea	33	2	1
China P.R.	32	32	26
Germany	32	31	31
Spain	28	26	25
Canada	23	21	24
Brazil	21	18	18
Italy	16	16	16
India	12	11	9
United Kingdom	10	10	10
Austria	9	9	9
Australia	7	7	7
Poland	7	7	6

Source: Organic Certification Directory 2007 (Grolink 2007)

Most certification bodies are located in the EU, USA, Japan, South Korea, China, Canada, and Brazil. Many of the listed certification organizations also operate outside their home country; most are based in a developed country, but also offer their certification services in developing countries. Very few operate in several developed countries. For example, not a single EU-based certification body offers its services in the US, including those that have the required NOP accreditation. A handful of certifiers work on several or all continents.

Seventy-four countries have a home-based certification organization. Most of Africa and large parts of Asia still lack local service providers. There are only eight certification bodies in Africa (in South Africa, Kenya, Uganda, Tanzania and Egypt). Asia has 147 certification bodies, and most are based in South Korea, China, India, and Japan.

---

[1] Gunnar Rundgren, Grolink, Höje, Sweden, www.grolink.se/
[2] The Organic Standard, www.organicstandard.com

Table 11: Number of certification bodies per region

	2007	2006	2005	2004	2003
Africa	8	8	7	9	7
Asia	147	93	117	91	83
Europe	172	160	157	142	130
Latin America & Caribbean	47	43	43	33	33
North America	83	80	84	97	101
Oceania	11	11	11	11	10
Total	468	395	420	383	364

Source: Organic Standard

Certification bodies per region

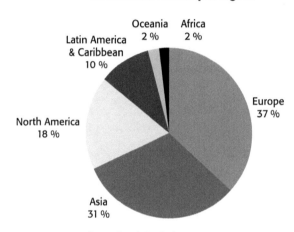

Source: Organic Standard 2007

Figure 16: Distribution of certification bodies by region 2007

Source: Organic Standard

Certification bodies were asked for information about the number of operators they certify. With 226 respondents, total number of operators came to 163'000. The number of farmers was specified by 213 certification bodies; in total, they certify 470'000 farms, with IMO's head office alone reporting more than 120'000 and its office in Latin America 32'000. BioLatina certified 20'000 farms, Naturland reports 46'000 farms, and Certimex 30'000 farms. It should be noted that the same farm may be certified twice; for example, many Naturland certified farmers are also IMO certified, as the two organizations cooperate closely. Nevertheless, the number of certified farms is likely to be at least 600'000, as data is lacking from Brazil, Japan, USA, South Korea, and China.

Most organizations are still not transparent about their turnover. Only 70 organizations responded. Many report figures in the range of 100'000 to 500'000 Euros. Bio Suisse and the Soil Association are heavy-weights when it comes to turnover, with more than 5 million Euros each. Other organizations reporting a high turnover are IMO, bio.inspecta, ICEA, CCPB, Suolo & Salute, Qualité France, DIO, Skal and Debio.

Of the 293 certification bodies that responded to the question concerning the starting date of their operation, only 11 started before 1985. Fifty-nine percent started after 1997.

Table 12: Number of certification bodies and approvals per region

Region	Total	IFOAM	Japan	ISO 65	EU	USA
Africa	8	0	0	4	0	0
Asia	147	6	34	10	11	12
Europe	172	15	12	75	148	33
Latin America & Caribbean	47	5	1	16	5	11
North America	83	6	10	25	0	64
Oceania	11	4	6	3	7	5
Total 2007	468	36	63	133	171	125
Total 2006	395	32	64	129	160	112
Total 2005	419	31	100	113	143	115
Total 2004	385	30	95	96	132	112
Total 2003	364	26	81	74	112	

There has been little movement in the approval status of organizations compared to 2006 (see table). Despite ISO 65 being claimed by some to be the universal standard for organic certification bodies, less than a third (133) have ISO 65 accreditation. The number of organizations approved in Japan (64) is stable. The EU has 171 approved bodies, with 23 non-EU-based bodies recognized within its system. The majority of imports into the EU come through certification granted under article 11.6 the EU regulation on organic farming commonly referred to as the 'importer's derogation.' Under that system, import authorizations were granted from 108 countries in 2005. The US system has 125 approved bodies, of which 71 are outside the USA.

Similar to the two previous years, only five organizations - four Italian and one from Australia - reported all five approvals.

75

Accreditation

Gerald A. Herrmann[1]

Organic agriculture is based on the commitment of farmers and processors to work according to set standards and regulations that define the organic production system. Transparency is an important component of organic production systems, ensuring that they are comprehensive and reliable, and as a result engender the consumer confidence necessary for market development. In this context, certification and accreditation systems serve primarily as tools to enhance trade, market development and foster confidence. Accreditation and certification mechanisms are developing rapidly. There are almost no spheres of human activity or technologies for which regulations or norms have not yet been developed and introduced. With regard to food, organic food production and processing set the precedents for the conventional industry.

Whereas private (farmer) organizations developed the standards for production, inspection and certification in the 1980s, the first governments took over this task at the beginning of the 1990s. Although they took on the task of defining the rules as a sovereign right, they did not necessarily become involved in the implementation of these rules at all levels.

Codex Alimentarius, with its organic chapter, defines the common international framework for governments (see chapter on standards and regulations). Regulations similar to those in the EU or US were passed and implemented at governmental or supra-governmental levels. State governments added specific requirements.

The private sector standards are based on the IFOAM Basic Standards, which were - and still are - a reference for governmental regulations and the Codex Alimentarius guidelines.

Today, about 80 countries are already regulating or are working towards regulating organic agriculture through national standards. Many countries are also developing additional requirements to regulate the recognition of inspection bodies; some are defining inspection procedures as well.

The major markets for imports and consumption, including Europe, the US and Japan are leading the way, but countries such as India, China and Brazil are following the same path. Inspection and certification is accredited or at least supervised by government authorities as defined in the regulations, even though the systems being implemented might be quite different. Control and supervision at all levels should guarantee that all inspectors and certifiers are evaluated and accredited (accreditation means 'the evaluation of certifiers'). The European Union has passed the revision of its organic regulation, although most of the details and technical requirements are still pending discussion and decision. The new import regulation seems to be more favorable for operators outside of the EU, including for certification bodies in Third Countries that may in the future be allowed to apply for accreditation by the European Commission directly. Procedural details, however, are pending.

---

[1]Organic Services GmbH, Landsberger Str. 527, D - 81241 Muenchen. Gerald A. Herrmann is the President of the International Federation of Organic Agriculture Movements IFOAM.

In addition, many private standards and labeling schemes exist, primarily in areas of the world where the organic market is well developed. Registers count more than 400 certification bodies certifying according to private standards and/or set regulations around the world (see article by Rundgren in this book). Private standards are also being initiated in less developed regions, aided through the support of international organizations and development agencies; IFOAM and UNCTAD have, for example, systematically supported the development of a private certification system for East Africa (see article by Sophia Twarog in this book).

However, it is not enough to define the rules. It is still necessary to achieve a minimum (worldwide) equivalency – guaranteed - throughout the system in order to let products flow. Lacking acceptance and recognition between different certification and accreditation systems contradicts the objective of enhancing trade, market development and fostering confidence. It is the opposite. The existing and different systems today create a technical barrier to trade that forces producers to seek multiple certifications for their operation if their products are to access different markets with different regulations.

Nobody can seriously state that this situation makes the organic system more safe and reliable. It is usually added bureaucracy without additional value, neither for the producer nor the consumer. Nonetheless, it is undisputed that the development of standards, certification and accreditation systems during the last decade has improved reliability and transparency of the organic system. From an overall perspective, though, it is now high time to consider how systems can be reviewed in order to make them more inclusive and to reduce market barriers. The differences in the organic systems more or less result from minor details and different (cultural) approaches, although all systems serve the common idea of improving credibility of the organic system.

In order to strengthen organics, all involved parties, like governments, private standard setting and certification bodies and other stakeholders, should concentrate their focus on the essential differences between organic and conventional, rather to struggling within the movement about differences, even details. It is high time to reconsider that accreditation and certification is basically a tool to strengthen the organic development. Because of the aforementioned factors, certification (including inspection and accreditation) should be reasonably designed to support the credibility of the organic system, rather than to endanger it by overburdening it with more and more details. This is what the organic movement is still trying to achieve with harmonized international basic standards and with designing a private system, yet acknowledging the reality of its practical restrictions.

It is yet to be seen whether the International Task Force on Harmonization and Equivalence in Organic Agriculture (ITF), a joint initiative of IFOAM, FAO and UNCTAD, will achieve consensus on harmonizing private with government and government with government standards/regulations. Both the private sector and representatives of governments are participating in this initiative (see following article by Sophia Twarog on the ITF).

Concerning the harmonization requirements for the accreditation of certifiers, the ITF recently initiated a project for the development of International Requirements for Organic Certification Bodies, (IROCB) taking both ISO 65 and IFOAM Criteria into account; these requirements are expected to be finalized and presented to the public in 2008.

The ITF – the FAO/IFOAM/UNCTAD International Task Force on Harmonization and Equivalence in Organic Agriculture[1]

SOPHIA TWAROG[2]

The Problem

Rapid growth of organic markets in developed countries presents promising opportunities for producers and exporters of organic products, including in developing countries. Discussions in a number of forums, including UNCTAD, FAO and IFOAM, have indicated that the plethora of certification requirements and regulations are considered to be a major, if not the key, obstacle for a continuous and rapid development of the organic sector, especially for producers in developing countries. The organic market is confronted with hundreds of private sector standards and governmental regulations, two international standards for organic agriculture (Codex Alimentarius Commission and IFOAM) and a host of conformity assessment and accreditation systems. Mutual recognition and equivalency among these systems is extremely limited. Lack of cooperation and 'harmony' is a central problem.

Food and Agriculture Organization of the United Nations

INTERNATIONAL FEDERATION OF ORGANIC AGRICULTURE MOVEMENTS

UNITED NATIONS
UNCTAD
United Nations Conference on Trade and Development

To service their clientele, certifying bodies must often obtain a number of costly accreditations. Producers interested in selling in more than one market, or even in more than one supermarket in a market, must bear the high costs of multiple certifications. In most markets, it is not enough to simply meet the government regulations; producers must also meet the specific requirements of the private organic standard, whose associated label is used by a particular retailer or recognized by consumers in that market. For developing country producers and exporters, this exercise may be even more costly, as they often required to bring in foreign inspectors to carry out the certification. Moreover, requirements in import markets may be poorly suited to the agro-ecological and socio-economic conditions of the producing countries. There are certainly many farmers around the world who would be interested in producing certified or-

---

[1] The text of this article is largely based upon ITF documents, including the ITF Communiqué from the sixth ITF meeting (September 2006, Stockholm) which is available in its entirety in Volume Four of the ITF Background Papers Series (UNCTAD/DITC/TED/2007/14).

[2] Dr. Sophia Twarog, Economic Affairs Officer, Trade, Environment and Development Branch, United Nations Conference on Trade and Development UNCTAD/DITC, E.8015, Palais des Nations, 1211 Geneva 10, Switzerland, www.unctad.org/trade_env!
Sophia Twarog is member of the ITF Steering Committee, which comprises representatives of the three convening organizations: Nadia El-Hage Scialabba, Food and Agriculture Organization of the United Nations (FAO), Selma Doyran, Joint FAO/WHO Food Standards Programme, Gunnar Rundgren and Antonio Compagnoni, International Federation of Organic Agriculture Movements (IFOAM) and Ulrich Hoffmann and Sophia Twarog (UNCTAD). Diane Bowen (IFOAM) is the ITF Secretariat Coordinator.

ganic products, but never do so in the face of such complicated and costly procedures.

Joining forces to seek solutions

Sharing great concern over the problems outlined above, IFOAM, FAO and UNCTAD decided in 2001 to join forces to search for solutions. The three organizations have complementary areas of competence that are all central to the problem: IFOAM in organic agriculture, FAO in agriculture in general and food security, and UNCTAD in trade and development.

The three organizations[1] agree that harmonization, mutual recognition and equivalence in the organic sector offer the only viable solution to the problem outlined above.

Accordingly, they jointly organized The Conference on International Harmonization and Equivalence in Organic Agriculture, in Nuremberg, Germany, 18-19 February 2002. This event was the first of its kind, where the partnership between the private organic community and United Nations institutions offered a forum for public and private discussions. The Conference was attended by 210 participants from 52 countries, including 42 government representatives, producers, certifiers, accreditors, traders, retailers and consumers involved in organic agriculture.

The International Task Force on Harmonization and Equivalence in Organic Agriculture

One of the key recommendations of the Conference was that 'a multi-stakeholder Task Force composed of representatives of Governments, FAO, UNCTAD and IFOAM should be established in order to elaborate practical proposals and solutions.'

In response, the International Task Force on Harmonization and Equivalence in Organic Agriculture (ITF) was launched at its first meeting on 18 February 2003 in Nürnberg, Germany. The ITF is an open-ended platform for dialogue between public and private institutions[2] (intergovernmental, governmental and civil society) involved in trade and regulatory activities in the organic sector. It aims to seek solutions to facilitate international trade in organic products and access of developing countries to international organic markets.

The ITF focuses on opportunities for harmonization, recognition, equivalence and other forms of cooperation within and between government and private sector organic guarantee systems. It commissions technical studies to fill information gaps and meets at least once a year to discuss and agree on next steps. It publishes the results of its work in books and on a dedicated website[3]. To date, the ITF has met seven times, most recently in Bali in November

---

[1] UNCTAD's Commission on Trade in Goods and Services, and Commodities, on its seventh session, in February 2002 asked the secretariat to 'examine ways to promote the practical application of the concept of international equivalence and mutual recognition, including between governmental and private sector standards.' (TD/B/COM.1/L.21)
[2] ITF participants have so far come from government agencies of 23 countries (Argentina, Australia, Brazil, Canada, China, Costa Rica, Denmark, Dominican Republic, Guatemala, Germany, Greece, India, Indonesia, Japan, Netherlands, Philippines, Sweden, Switzerland, Tanzania, Thailand, Tunisia, Uganda and USA), eight inter-governmental agencies (EU, OECD, FAO, SPC, UNCTAD, UNECE, UNEP and WTO) and 19 civil society organizations and businesses (Argencert, Ecocert, Ecologica, Green Net, IFOAM, IAF, ISF, IOAS, ISEAL Alliance, ICEA, JONA, Kawacom Uganda Ltd., KOAN, KRAV, Migros, OFDC, Oregon Tilth, Rachel's Organic Dairy and Women in Business Development)
[3] www.unctad.org/trade_env/itf-organic

2007, and published four volumes of ITF Background papers.[1]

ITF work is divided into three overlapping phases: Review Phase, Solutions Phase, and Communications Phase.

The Review Phase of the ITF work (2003-2005) analyzed the impact of existing organic regulations on trade, current models and mechanisms that enable organic trade, experiences of cooperation, recognition and equivalence in the organic sector, and potential models and mechanisms for harmonization, equivalence and mutual recognition.

The current Solutions Phase of the ITF agreed to pursue a strategy comprised of the following elements:

- a single international reference standard for organic production, as a basis for regional and national standards;
- a mechanism for the judgment of equivalence, based on the reference standard;
- one international requirement for organic certification bodies; and
- common international approaches for recognition or approval of certification bodies.

The ITF also agreed to:

- use or adapt existing structures and mechanisms of regulation, rather than establishing new entities;
- give special consideration to the situation of developing countries; and
- gear actions towards cooperation at and between all levels: among and between governments (with or without an organic regulation), accreditation bodies and certification bodies.

The ITF agreed that solutions should support the continued growth of organic agriculture and maintain its principles. They should fulfill the additional criteria of: benefits to both producers and consumers; respect for national sovereignty; access to all markets with minimal bureaucracy; fair competition; adherence to fair trade practices; consumer protection; context-sensitivity; stakeholder support and participation; market choice; and transparency.

The ITF is currently developing the following tools:

- a set of essential International Requirements for Organic Certification Bodies (IROCB), as a basis for equivalence and future harmonization; and
- a guidance document for judging equivalency of organic standards (Equitool).

In light of the progress achieved so far, the ITF has recognized that a single reference for organic standards is not yet a feasible proposition; although the guidelines of the Codex Alimentarius Commission (CAC) and IFOAM Basic Standards (IBS) are very similar in content, their scope and governance are too distinct to be merged. The ITF however realized that having two international reference standards, from the public and private sector re-

---

[1] Please see UNCTAD-FAO-IFOAM (2005, 2006, 2007, 2008) Volumes 1-4 of the Background papers of the International Task Force on Harmonization and Equivalence in Organic Agriculture (United Nations document symbols UNCTAD/DITC/TED/2005/4, UNCTAD/DITC/TED/2005/15, UNCTAD/DITC/TED/2007/1, UNCTAD/DITC/TED/2007/14). These can be downloaded from the ITF Website at www.unctad.org/trade_env/itf-organic/publications1.asp.

spectively, is valuable, provided that there is effective linkage between the sectors.

The ITF recommends that:

- Countries make every effort to utilize the ITF results in order to facilitate trade, and in their efforts to build or enhance the organic sector;
- Public-private participation be improved in decision-making for both international organic standards (i.e. CAC and IBS);
- Governments commit to using international standards as the reference point for import approvals;
- The International Requirements for Organic Certification Bodies, being developed by the ITF on the basis of ISO65 and the IFOAM Accreditation Criteria, be used when regulating imports and developing requirements for organic certification bodies;
- Governments and private accreditation systems develop mutual recognition, which will be based on the International Requirements for Organic Certification Bodies (IROCB);
- Equivalence of organic standards and technical regulations will be based on one set of criteria, which is being developed by the ITF (Equitool); and
- Consideration is given to emerging alternatives to third party certification, such as Participatory Guarantee Systems.

The ITF has now also commenced the Communications Phase of its work. Its goals are to raise awareness of and mobilize political support for its recommendations. ITF members and ITF Steering Committee members are making a concerted effort to promote global pick-up of the ITF body of knowledge and solutions, garnered from four years of intense technical analysis and animated and informed public-private sector experts dialogues.

Certainly the ITF has already had an impact in the world. The need for harmonization and equivalence is well recognized now. The new EU regulation governing imports is much more import-friendly, including for developing country products. Private certifying bodies have increased cooperation, and dropped many of their additional particular requirements in the context of mutual recognition agreements, such as those among IFOAM-accredited certification bodies.

The ITF now seeks participation - and engagement – by all in the ITF solutions and approach. The main message is to accept, in the context of trade, effective organic standards and conformity assessment that:

(a) are tailored to the agro-ecological and socio-economic conditions of the producing country;
(b) are equivalent to international standards for organic production either Codex or IFOAM basic standards (The Equitool being developed by the ITF could be useful in this regard); and
(c) meet international standards for organic conformity assessment (IROCB developed by the ITF).

In short, expect the best, but don't force the entire world to be like you.

Organic Aquaculture

STEFAN BERGLEITER[1]

Terminology

The term aquaculture implies the husbandry of aquatic organisms, and consequently repre-sents a subject as wide and as multifaceted as agriculture. A remarkable diversity of plant and animal species are cultivated; humans breed approximately 465 species from 28 differ-ent families of plants and 107 families of animals ranging from microscopic, unicellular blue-green algae (*Spirulina*), fragile shrimp (*Litopenaeus*, *Macrobrachium*) and hard-shelled oysters (*Ostrea*, *Crassostrea*) to sedate carp (*Cyprinus*), colorful ornamental fish (*neon tetra Paracheirodon*), and elegant salmon (*Salmo*, *Oncorhynchus*). Surprisingly, even huge tuna (*Thunnus*) and multi-meter long kelp (*Laminaria*, *Undaria*)[2] are counted among cultivated species.

Unsurprisingly, this leads to a similar variety in aquaculture techniques that vary according to the social, cultural, and ecological context of production.

Emergence and legal framework

Certified Organic Aquaculture is a quite recent initiative. From the mid 1990s onwards, a number of certification agencies and organic growers' associations[3] began developing spe-cific aquaculture standards. In Germany and Austria, organic farmers who were running carp ponds as a source of additional income served as the main trigger for this process; they demanded adequate quality management and certification for this production.

In the UK and Scandinavia, the impetus for the development of organic aquaculture stems from increasing pressure on conventional salmon farms by environmental organizations. This public pressure motivated the industry to consider a more ecologically sustainable production system.

Given the fact that organic aquaculture is in relatively early stages of development, the legislative and institutional framework is also just beginning to take shape. In 1999, when the EU Regulation on organic animal husbandry (Council Regulation (EC) No 1804/19999) was published as a supplement to organic production, aquaculture was not covered, and was explicitly left to private and national certification initiatives. Currently, the European Com-mission is drafting an organic aquaculture regulation, with the aim of regulating organic aquaculture in all EU countries. The US National Organic Program (NOP), the national regulation there that came into force in 2002, also does not contain any provisions for organic aquaculture. However, in a process comparable to that of the EU, policies regulating

---

[1] Dr. Stefan Bergleiter, Naturland e.V., Kleinhaderner Weg 1, 82166 Gräfelfing, Germany, www.naturland.de
[2] In Southeast Asia, there are farms producing frogs (*Rana*), crocodiles (*Crocodylus*) or soft-shell turtles. Even if these may be an economically attractive widening of the product range, they often appear problematic from an animal welfare point of view.
1010[3] DEBIO/Norway, Ernte/Austria, KRAV/Sweden, Naturland/Germany and the Soil Association/UK

organic aquaculture shall soon be integrated into the existing standard. Both processes are anticipated to be finalized by 2009.

Providing guidance to private certification systems and serving as a global communications platform, the Aquaculture Group of the International Federation of Organic Agriculture Movement aims to serve this specialized area.

Availability of organic aquaculture products

As in other segments of the organic sector, market demand is the driving force for activities at the producer level, and also for the conversion of aquaculture companies to certified organic management.

A stable supply, attractive both in terms of quality and price, is necessary for generating interest by wholesale and retail markets in a new product, which is particularly true for organic aquaculture. While other organic food items reach their consumers through rather diverse avenues, such as farms stores, farmers' markets and box schemes, aquatic products generally require a sophisticated logistical infrastructure, especially concerning cooling systems from processing to consumption. The increasing number of refrigerated counters in natural food stores and a stronger commitment by conventional supermarkets has had a very positive impact on the development of fresh organic aquaculture sales.

Consultancy in an organic shrimp project in Vietnam. The picture shows the mangrove nursery

Picture: Stefan Bergleiter, Naturland, Germany

Since the mid 1990s, the progression of certified organic aquaculture has been characterized by a steady increase of product volumes on the market. Organic salmon projects initiated by Naturland and the Soil Association in Ireland and Scotland have led this evolution. Gradually, organic aquaculture lost its image as a purely niche activity, and bigger retail companies in Germany, the United Kingdom and Switzerland added aquaculture products into their assortment, which in turn encouraged more producers in many countries to convert to organic production. The table below offers an overview of certified organic aquaculture products available.

Table: Available products from certified organic aquaculture (December 2007)

Species	Country of Origin[1]	Type of Processing
Carp with accompanying species (tench, grass carp, pike perch, eel, European catfish etc.)	Germany, Austria, Hungary	Mainly fresh, some smoked, spreads
Brown trout, rainbow trout, (mainly hybrid) char	UK, Germany, Italy, France, Spain, Denmark, Norway, Sweden, Switzerland, Austria	Fresh, smoked, spreads.
Mediterranean sturgeon	Spain	Fresh, caviar
Atlantic salmon	UK, Ireland, Norway	Fresh, frozen, smoked, marinated
Sea bass (Loup de Mer), Sea bream (Dorade), Drum species	France, Greece, Israel	Fresh, frozen whole fish
Tilapia	Honduras, Ecuador, Israel, Brazil	Frozen, filets
Pangasius (Mekong Catfish)	Vietnam	Frozen filets
Green shell mussel (Perna canaliculus)	New Zealand	Frozen
Shrimp (Western White Shrimp)	Ecuador, Peru, Brazil	Frozen (tails)
Shrimp (Black Tiger Shrimp)	Vietnam, Java (Indonesia), Thailand	Frozen (tails)
Micro algae (Spirulina, Chlorella)	Taiwan, Hawaii (USA), India, China	Powder, tablets, tea

The principles of organic aquaculture – Development of standards

As in other areas of organic food production, organic associations and certification bodies must regulate important technical aspects of organic aquaculture through relevant standards. In general, these standards represent the result of dialogue between industry, science, environmental organizations, organic associations, and other stakeholders. The standards represent dynamic formulations that are modified as necessary to address changes in the industry, and to incorporate scientific insights or to take into account new technical developments. Keeping the flexibility to enable changes to the fledgling standards of aquaculture is particularly important, as the sector is still rapidly developing, and organizations have dealt with the related issues for a short period.

---

[1] The list refers to farming operations that have undergone a regular certification procedure. This means that the certifying body is known and that the standards are publicly available.

Polyculture (pigs, fruitbearing trees, mangrove trees) on an organic shrimp farm in Ecudador

Picture: Stefan Bergleiter, Naturland, Germany

The process of developing organic aquaculture standards globally should take place in a harmonized way, while ensuring the basic principles of organic farming are respected. Fortunately, many organic certification agencies adhere to the IFOAM Basic Standards on a voluntary basis or in the context of accreditation with IFOAM. The IFOAM Basic Standards are designed to grant sufficient flexibility to certification agencies and to permit specific local circumstances to be addressed, while holding firm positions on fundamental and universal requirements, such as the prohibition of genetic engineering.

Necessarily, certification standards differ significantly between the various aquatic organisms,[1] but they refer to common principles:

-    The production systems must correspond to the species-specific physiological and behavioral needs of animals, which means the installation of suitable substrates and hiding places, and – in most cases – limitation of stocking densities.

-    Feedstuffs must be either from certified organic agriculture or from suitable fishery products. Synthetic amino acids, synthetic pigments, and fishmeal or oils from fisheries dedicated to fishmeal harvesting are not allowed.

---

[1] Salmon, carp, mussels shrimp and other organisms are cultivated according to organism specific standards, see www.naturland.de for more information.

- Use of chemical and synthetic pesticides and similar substances on the farm is prohibited. Net cages may not be treated anti-fouling agents to regulate algal growth.

- Prophylactic use of antibiotics and other conventional treatments is prohibited.

- Aquaculture operations may not yield any negative impact on neighboring eco-systems, which means effluent water quality and important natural vegetation, such as mangrove forests, must be thoroughly monitored.

- Absolute prohibition of GMOs (genetically modified organisms), referring not only to feed and other inputs, but to the animal stock being cultivated.

- Processing takes place according to organic principles, which prohibit the use of sodium metabisulfite in shrimp processing, application of liquid smoke in salmon smokeries and prohibit the use of many other conventional processing aids.

- Social sustainability is an increasingly important aspect of organic aquaculture. In addition to general requirements for social responsibility in employment relationships, it must be taken into consideration that other resource users are not affected negatively by the aquaculture operation, such as fishermen in the vicinity of a shrimp farm losing access to traditional fishing grounds.

AFRICA

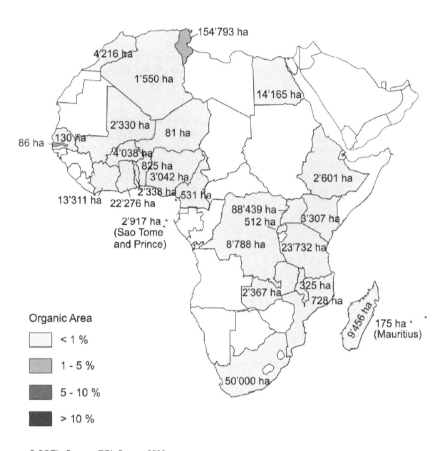

154'793 ha

4'216 ha

1'550 ha

14'165 ha

2'330 ha

81 ha

86 ha

130 ha

4'038 ha

825 ha

3'042 ha

2'601 ha

2'338 ha 531 ha

13'311 ha 22'276 ha

88'439 ha

512 ha

3'307 ha

2'917 ha
(Sao Tome
and Prince)

8'788 ha

23'732 ha

325 ha

2'367 ha

728 ha

9'456 ha

175 ha
(Mauritius)

50'000 ha

Organic Area

☐ < 1 %

▨ 1 - 5 %

▨ 5 - 10 %

▨ > 10 %

© SOEL, Source: FiBL Survey 2008

Map 2: Organic farming in Africa 2006

Organic Farming in Africa[1]

HERVÉ BOUAGNIMBECK[2]

Introduction

Certified organic production is mostly geared to products destined for export markets. However, local markets for certified organic products are growing, especially in Egypt, South Africa, Uganda, Tanzania and Kenya.

It is difficult to get a clear sense of the scale of organic production – referring to both certified and non-certified organic production – in Africa. Data related to organics in Africa has only begun to be collected in recent years, and as such, figures are sometimes approximate and incomplete.

Organic agriculture appears to be a viable and sustainable development option, particularly for (groups of) smallholder farmers in Africa (Lyons and Burch, 2007). Local NGOs and farmers' groups and development agencies are increasingly adopting organic techniques as a method of improving productivity and addressing the pressing problems of food security and climate change. Organic farming fits in a package that addresses a number of concerns. They resonate with, and are being used in, initiatives designed to:

- maintain and enhance soil fertility;
- mitigate the effects of climate change and reverse desertification;
- promote tree-planting and agro forestry;
- maintain and improve biodiversity;
- promote the use of local seed varieties and develop strategies for combating pests based on locally available inputs;
- strengthen group formation, promote joint marketing;
- support the most vulnerable social groups such as female headed households;
- improve human health by avoiding chemical use and diversifying diets;
- ensure food security; and
- contribute to the eradication of poverty.

Certified organic land

Statistics for certified production are provided in the tables at the end of this chapter. These statistics are probably incomplete - most countries do not have data collection systems for organic farming. Data on the numbers of farms are not available for every country.

---

[1] This article represents a totally revised and updated version of the article by Elzakker et al., published in the 2007 edition of 'The World of Organic Agriculture.'

[2] Hervé Bouagnimbeck, IFOAM Africa Office, c/o IFOAM Head Office, Charles-de-Gaulle-Str. 5, 53113 Bonn, Germany, E-mail h.bouagnimbeck@ifoam.org, Tel. +49 228 926 50-10, Fax +49 228 926 50-99, Internet www.ifoam.org/about_ifoam/around_world/africa.html

According to these figures, 417'059 hectares are currently managed and certified organic by at least 175'266 farms; the countries with the largest organically managed land are Tunisia (154'793 hectares), Uganda (88'439 hectares), South Africa (50'000 hectares) and Tanzania (23'732 hectares). Additionally, 8.3 million hectares are certified as forest and 'wild harvested' areas. The largest wild collection areas are in Zambia (7.2 million hectares), Sudan (490'000 hectares), Kenya (186'000 hectares) and Uganda (158'328 hectares).

Markets

Farmers in Africa produce a diversity of organic crops. The range of certified organic products currently being produced in Africa is listed below.

Organic Produce from Africa (by type and country)

Product Group	Countries
Bananas	Mali, Cameroon, Ghana, Rwanda, Senegal, Uganda
Cereals incl. Rice	Egypt, Ethiopia, Madagascar, Mozambique, Sudan
Citrus Fruits, Grapes (including wine)	Egypt, Morocco, South Africa
Cocoa	Cameroon, Ghana, Madagascar, Tanzania, Uganda, Sao Tome and Prince
Coconut Oil	Mozambique
Coffee	Cameroon, Ethiopia, Kenya, Madagascar, Rwanda, Tanzania, Uganda
Cotton	Benin, Burkina Faso, Egypt, Mali, Senegal, Sudan, Tanzania, Uganda, Zambia
Dried Fruits	Algeria, Benin, Burkina Faso, Cameroon, Egypt, Ghana, Madagascar, Morocco, Tanzania, Togo, Tunisia, Uganda
Essential Oils	Ghana, Kenya, Madagascar, Tanzania, Uganda, Zimbabwe
Fresh Vegetables	Cameroon, Gambia, Egypt, Kenya, Madagascar, Malawi, Mali, Morocco, Rwanda, Sao Tome and Prince, South Africa, Tunisia, Zambia
Ground Nuts (peanuts)	Cameroon, Mozambique, Tanzania, Zambia
Gum arabic	Chad
Herbs (culinary)	Egypt, Ethiopia, Ghana, Kenya, Madagascar, Malawi, Morocco, Mozambique, South Africa, Tunisia, Zambia, Zimbabwe
Honey	Algeria, Malawi, Tanzania, Tunisia, Uganda, Zambia
Medicinal / Therapeutic Herbs and Spices	Egypt, Morocco, Namibia, Tunisia, Zambia
Olive Oil	Tunisia
Other tropical fresh fruits	Cameroon, Egypt, Ghana, Madagascar, Senegal, South Africa, Tanzania, Uganda
Palm Oil	Ghana, Madagascar
Processed fruits incl. Juices	Ghana, Tanzania, Uganda
Sesame	Burkina Faso, Mali, Mozambique, Senegal, Uganda, Tanzania
Spices (culinary)	Cameroon, Egypt, Ethiopia, Madagascar, Malawi, Mozambique, South Africa, Tanzania, Uganda, Zimbabwe
Sugar	Cameroon, Madagascar, Mauritius,
Tea	Kenya, Tanzania, Rwanda
Tree Nuts (cashew, shea)	Burkina Faso, Ghana, Kenya, Malawi, Mali, Morocco, Tanzania, Togo, Uganda

Source: Parrot 2006, with updates from the IFOAM /FiBL Survey 2008

Exports

The majority of certified organic produce is destined for export markets, with the large majority being exported to the European Union, which is Africa's largest market for agricultural produce. Export of organic produce from Egypt totaled 12'542 tons in 2004; this figure represented some 90 percent of organic production (ECOA, 2005). Export of organic produce from Uganda, as summarized in the table below, totaled 7'877 tons in 2006. Data on current volume of produce are not available.

Export volume for Uganda in 2006

Product	Export volume in tons
Fruits	835
Cotton	3'875
Sesame	1'124
Coffee	1'705
Shea	1
Vanilla	35
Cocoa	280
Barkcloth	1
Fish	3
Hibiscus	15
Chilly Pepper	5
TOTAL	7'877

Source: NOGAMU/IFOAM-FiBL Survey 2008

The domestic market

The African market for organic products is still small. Certified organic products are currently recognized in only a few domestic markets, including Egypt, South Africa, Uganda, Kenya and Tanzania (Rundgren and Lustig, 2007). There are a number of factors expanding the domestic market in these countries. To begin, there is a growing middle class – most notably in Egypt and South Africa – who share some similar values to European organic consumers. As a result, there are growing domestic market opportunities for diverse organic products in these locations, including organic fresh fruit and vegetables, dairy products, meat, wine herbs and beauty products.

In Egypt, mainly in Cairo, specialized shops and a number of supermarket chains (Metro and Carrefour) have organic sections, and sell mostly fruits and vegetables. Similarly, organic shops in South Africa and Uganda have also raised the profile of organic produce. In South Africa, the major supermarkets stock organic products. The NOGAMU shop in Kampala, Uganda, is directly adjacent to the NOGAMU office. Visitors to the shop have direct access to information about organic agriculture and related issues (Lyons and Burch, 2007).

State support, standards and legislation

Organic agriculture is mainly not integrated into state agricultural policies. In some countries, mostly in East Africa, policy development is being undertaken, and the national organic movements are strongly involved in the process (Elzakker et al. 2007).

For exports, most African countries rely upon foreign standards. To date, the majority of organic production that is certified in Africa has been certified according to the EU regulation for organic products. Some producers are also certified to the US National Organic Program or the organic standards of the Japan Agriculture Standards (JAS) and to numerous private-sector organic standards, such as Soil Association, KRAV and Naturland (Rundgren, 2007).

Certification services have been offered by foreign-based certification bodies. Originally, all work, including inspection, was carried out by foreigners flying in and out. Gradually, more and more work (in particular the inspections) is done by local staff. Some certification bodies have established regional representation or developed closer cooperation with national bodies (Elzakker et al. 2007).

Costs of certification are generally considered to be high. In East Africa, for example, an individual farm will pay 500 to 3'000 US Dollars for a foreign certification. However, the cost per individual farmer in an Internal Control System (ICS) can be as low as a few US Dollars for very big groups. For a typical ICS with 500 farmers, the cost is likely to be in the range of 10 US Dollars per farmer, and for very small ICS groups maybe as high as 100 US Dollars per farm. There are also substantial costs involved in the operation of the ICS itself (Rundgren, 2007).

Launch of the East African Organic Products Standard and the East African Organic Mark

The East African Organic Products Standard (EAOPS) - the second regional organic standard in the world, following that developed by the European Union - and the associated East African Organic Mark (EAOM) were developed by a public-private sector partnership in East Africa, supported by IFOAM, and the UNCTAD-UNEP Capacity Building Task Force on Trade, Environment and Development (CBTF), a joint initiative of the United Nations Conference on Trade and Development and the United Nations Environment Program. Both the mark and the standard were officially launched by the Prime Minister of Tanzania at the East African Organic Conference in May 2007 (see chapter by Twarog on the East African Organic Standards).

Linked to the development of the EAOPS, various consumer awareness-raising activities were implemented to increase consumer awareness in East Africa and support the introduction of the organic mark.

The EAOPS serves to unite organic regulations and markets in Kenya, Tanzania, Uganda, Rwanda, and Burundi. It is expected that the organic mark and the consumer awareness campaign will contribute to the development of the regional market.

The NGO sector

In several African countries, organic agriculture has reached a critical stage of development, and the national organic sectors have established national organic agriculture networks to represent the organic sector both national and international levels. These umbrella organizations serve to link the stakeholders of national movements, strengthen the sector and enhances its impact (Rundgren, 2007). The following national movements have been established:

- The National Organic Agricultural Movement of Uganda (NOGAMU);
- Organic Producers and Processors Association of Zambia (OPPAZ);
- Organics South Africa (OSA);
- The Organic Movement of Madagascar;
- The Kenyan Organic Agriculture Network (KOAN);
- The Organic Movement of Mali (MOBIOM);
- The Ghana Organic Agriculture Network (GOAN);
- The Ethiopian Association of Organic Agriculture (EAOA);
- The Tanzania Organic Agriculture Movement (TOAM); and
- The Zimbabwe Organic Producers' and Processors' Association (ZOPPA).

Research, extension and training

Agricultural research in Africa is fragmented between the international research centers, universities, national research institutes, and formal or informal field level research. There are some outstanding examples of innovative organic research at all of these levels. Pioneering research on organic farming techniques has emerged from the World Agroforestry Centre (formerly ICRAF) and the International Centre for Insect Physiology and Ecology (ICIPE). Other centers, such as the International Institute for Tropical Agriculture (IITA) and the International Livestock Research Institute (ILRI) could potentially contribute to finding solutions to the problems facing organic farmers (Elzakker et al. 2007). National initiatives, like that of the Uganda Martyrs University, the Organic Agriculture Project in Tertiary Institutions in Nigeria (OAPTIN), and the Olusegun Obasanjo Center for Organic Research and Development (OOCORD), are also growing.

The OOCORD was launched in 2007 and is based in the city of Ibadan. The Center aims to develop research and knowledge exchange on sustainable, organic agricultural systems to address the dual needs of food security and incomes in Africa. Collaboration and information exchange with other institutions is expected to strengthen OOCORD's functions (OOCORD, 2007).

Outlook

The fact that traditional African agriculture is low external input provides a potential basis for organic agriculture as a development option for Africa. Organic farming practices deliberately integrate traditional farming methods and make use of locally available resources. As such, they are highly relevant to a majority of African farmers.

There is undoubtedly room for a substantial increase in certified organic production in Africa, and smallholders engaged in it often derive significant benefits, improving their incomes, nutritional status and livelihoods as a result. Yet, there are also significant constraints on the potential for development. In part, these are external, to do with the costs of certification, problems of infrastructure, maintaining links with distant markets and the vagaries of world markets. There are internal constraints as well. The overarching priority for African agriculture is that of achieving sustainable food security. Organic agriculture has a huge potential in helping meet this aim.

In addition to expanding international market access, there is a need to develop local and regional markets for organic produce in Africa. Increased consumer awareness, cooperation among stakeholders and producers in the supply chain and the development of conformity assessment mechanisms for local marketing that are accessible for smallholders, such as Participatory Guarantee Systems (PGS), are key elements to achieving long-term sustainability of organic production systems.

Background: The IFOAM Africa Office

The International Federation of Organic Agriculture Movements (IFOAM) established the IFOAM Africa Office in 2004 to help the growth of organic agriculture on the continent. The Africa Office is presently based at the IFOAM Head Office in Bonn, Germany.

The IFOAM Africa Office coordinates IFOAM's advocacy network in Africa to take a proactive approach to the development of the organic sector. For the entire African continent, the IFOAM Africa Office empowers organic agricultural advocates - IFOAM Contact Points and other member organizations - to make concerted efforts to promote organic agriculture among farmer groups, NGOs, governments, and development organizations.

The Africa Office works through IFOAM Contact Points across Sub-Saharan Africa. Each of them represents or is the coordinating office for a national organic agriculture movement or of a national or regional organic network.

-   Ethiopian Association of Organic Agriculture (EAOA), Ethiopia
Contact: Ferede Addisu Alemayehu, E-mail: alfrd05@yahoo.com
-   Ghana Organic Agriculture Movement (GOAN), Ghana
Contact: Samuel Adimado, E-mail: adimadosam@yahoo.com
-   Kenya Organic Agriculture Network (KOAN), Kenya
Contact: Wanjiru Kamau, E-mail: wanjiruk@elci.org
-   Laulanié Green University and Association (LGU and LGA), Madagascar
Contact: Andrianjaka Rajaonarison, E-mail: njakar@gmail.com

- National Organic Agriculture Movement of Uganda (NOGAMU), Uganda
Contact: Susan Nansimbi, E-mail: snansimbi@nogamu.org.ug

- Organic Producers and Processors Association of Zambia (OPPAZ), Zambia
Contact: Munshimbwe Chitalu, E-mail: munshimbwe_chitalu@yahoo.co.uk

- Organic Agriculture Project for Tertiary Institutions in Nigeria (OAPTIN), Nigeria
Contact: Dr. Olugbenga Adeoluwa, E-mail: adeoluwaoo@yahoo.com

- Participatory Ecological Land Use Management (PELUM) Regional desk, Zambia
Contact: Marjorie Chonya, E-mail: inforunit@pelum.org.zm

- Sustainable Agriculture Development Network (REDAD), Benin
Contact: Abel Sekpe, E-mail: sulpa74@yahoo.fr

- Tanzania Organic Agriculture Movement (TOAM), Tanzania
Contact: Noel Kwai, E-mail: noelkwai2003@yahoo.com

- Zimbabwe Organic Producers' and Processors' Association (ZOPPA), Zimbabwe
Contact: Fortunate Nyakanda, E-mail: gadzirayichris@yahoo.co.uk

The Africa Office publishes the Africa Organic News, an electronic newsletter, on a monthly basis. The newsletter features news about organic agriculture in Africa, and it is distributed freely in English and French to a wide audience in and outside Africa in a format that can be printed and distributed locally. The newsletters are available at the Africa Office homepage at www.ifoam.org/newsletter/newsletter_africa/AOSC_Newsletter_Archive.html.

- Contact: Hervé Bouagnimbeck, IFOAM Africa Office, c/o IFOAM Head Office, Charles-de-Gaulle-Str. 5, 53113 Bonn, Germany, E-mail h.bouagnimbeck@ifoam.org, Tel. +49 228 926 50-10, Fax +49 228 926 50-99, Internet www.ifoam.org/about_ifoam/around_world/africa.html

References

ECOA (2005). Background Paper, Egyptian Center for Organic Agriculture, Cairo.

Elzakker, B. van, Parrott, N., Chola Chonya, M. and Adimado, S. (2007). Organic Farming in Africa. In: Willer/Yussefi (Eds.) (2007): The World of Organic Agriculture. Statistics and Emerging Trends 2007. International Federation of Organic Agriculture IFOAM, Bonn, Germany and Research Institute of Organic Agriculture (FiBL), Switzerland, pp. 93-105.

Lyons, K. and Burch, D. (2007). Socio-economic effects of Organic Agriculture in Africa, Germany, International Federation of Organic Agriculture Movements.

OOCORD (2007). Quarterly publication of OOCORD., Nigeria.

Parrott, N. and Elzakker, B. van (2003). Organic and like minded movements in Africa, Germany, International Federation of Organic Agriculture Movements IFOAM.

Rundgren, G. (2007). Participatory Guarantee Systems in East Africa, Germany, International Federation of Organic Agriculture Movements. Archived at http://www.ifoam.org/about_ifoam/standards/pgs.html

Rundgren, G. and Lustig, P. (2007). Organic Markets in Africa, Germany, International Federation of Organic Agriculture Movements, Bonn

Africa: Organic Farming Statistics

Table 13: Organic agricultural land and farms in Africa 2006

Country	Year	Organic agricultural area (ha)	Share of agricultural land	Farms
Algeria	2006	1'550	0.00%	39 (2005)
Benin	2006	825	0.02%	1'132
Burkina Faso	2006	4'038	0.04%	6'195
Cameroon	2006	531	0.01%	102
Chad[1]	2006		0.00%	36
Congo (Democr. Rep.)	2006	8'788	0.04%	5'150
Egypt	2006	14'165	0.41%	460
Ethiopia	2006	2'601	0.01%	784
Gambia	2006	86	0.01%	No data
Ghana	2006	22'276	0.15%	3'000
Ivory Coast	2006	13'311	0.07%	No data
Kenya	2006	3'307	0.01%	18'056
Madagascar	2006	9'456	0.03%	5'455
Malawi	2002	325	0.01%	13
Mali	2006	2'330	0.01%	5'840
Mauritius	2006	175	0.15%	5
Morocco	2006	4'216	0.01%	No data
Mozambique	2006	728	0.00%	1'928
Niger	2006	81	0.00%	No data
Nigeria	2006	3'042	0.00%	No data
Rwanda	2006	512	0.03%	20
Sao Tome and Prince	2006	2'917	5.21%	1'291
Senegal	2006	130	0.00%	1'020
South Africa	2005	50'000	0.05%	No data
Tanzania	2006	23'732	0.05%	22'301
Togo	2006	2'338	0.06%	5'101
Tunisia	2006	154'793	1.58%	862
Uganda	2006	88'439	0.71%	86'952
Zambia	2006	2'367	0.01%	9'524
Total		417'059	0.05%	175'266

Source: IFOAM/FiBL Survey 2008

---

[1] Chad has wild collection activities. The number of farms refers to these activities.

Table 14: Use of organic agricultural land in Africa 2006

Main land use type	Crop category	Agricultural area (ha)
Arable land	Cereals	212
	Fallow land as part of crop rotation	50
	Flowers and ornamental plants	30
	Medicinal & aromatic plants	8'062
	Oilseeds	9'885
	Other arable crops	4'752
	Protein crops	74
	Root crops	58
	Textile fibers	9'121
	Vegetables	1'955
Arable land		*34'182*
Cropland, other/no details		*15'620*
Permanent crops	Citrus fruit	3'981
	Cocoa	7'039
	Coffee	22'925
	Fruit and nuts	8'398
	Medicinal & aromatic plants	5'502
	Olives	89'324
	Other permanent crops	15'459
	Permanent crops, no details	1'817
	Sugarcane	180
	Tea	164
	Tropical fruit and nuts	8'657
Permanent crops		*163'447*
Permanent grassland		*50'305*
No information		*153'347*
Other		150
Total		417'059

Source: IFOAM/FiBL Survey 2008.

Table 15: Certified wild collection area in Africa 2006

Country	Crops harvested	Organic wild collection (ha)
Algeria	Fruit and berries	850
Burkina Faso	Nuts	9'887
Cameroon	Forest products	35
Chad	Gum arabic	83'400
Kenya	Bee pastures / Honey	86'028
	Wild, collection, no details	100'000
Madagascar	Wild collection, no details	17'302
Mali	Nuts	458
Morocco	Argan oil	100'000
Niger	Wild collection, no details	50
Nigeria	Forest products	100
South Africa	Wild collection, no details	1'409

Country	Crops harvested	Organic wild collection (ha)
Sudan	Gum Arabic	490'000
Tanzania	Bee pastures	3'077
Tunisia	Forest products	65'683
Uganda	Wild collection, no details	158'328
Zambia	Bee pastures	7'000'000
	Wild collection, no details	185'075
Total		8'301'682

Source: IFOAM/FiBL Survey 2008

Data and information sources/contacts for the IFOAM/FiBL Survey

- Algeria: Data provided by / Source: Dr. Lina Al Bitar, MOAN, IAMB, Bari, Italy
- Benin: Data provided by / Source: Abdoul Aziz Yanogo, ECOCERT SA West Africa Office
- Burkina Faso: Data provided by/Source: Ecocert, Ouagadougou, Burkina Faso
- Cameroon: Data provided by / Source: Guy Wamba and Lazare Youmbi, Ecocert, Douala, Cameroon
- Chad: Only Wild collection area; data provided by Abdoul Aziz Yanogo, ECOCERT Ouagadougou: Source: Abdoul Aziz Yanogo, ECOCERT Ouagadougou
- Egypt: Data provided by Paul Kledal,Institute of Food and Resource Economics, University of Copenhagen, Denmark, based on data of the certifiers
- Ethiopia: Certifier data
- Gambia: Data provided by / Source: Ecocert, Ouagadougou, Burkina Faso
- Ghana: Data provided by / Source: Samuel Adimado, Ghana Organic Agriculture Network, Kumasi, Ghana
- Ivory Coast: Data provided by / Source: Noufou Coulibaly, Organization Professionelles Agricoles, Ivory Coast; furthermore certifier data.
- Kenya: Data provided by / Source: Jack Muga Juma, Kenya Agriculture Organic Network, KOAN Secretariat, Nairobi, Kenya
- Madagascar: Data provided by Andrianjaka Rajaonarison, Laulanié Green Association (LGA), Antananarivo, Madagascar; Source: Sandra Randrianarisoa, Ecocert Madagascar/East Africa, Antananarivo, Madagascar
- Mali: Data provided by / Source: Andrea Bischof, Helvetas, Bamako, Mali and Ecocert, Ouagadougou, Burkina Faso
- Mauritius: Data provided by / Source: Sunita Facknath and Bhanooduth Lalljee, Faculty of Agriculture – University of Mauritius, Reduit, Mauritius
- Morocco: Data provided by Professor Lachen Kenny, Institut Agronomique et Vétérinaire Hassan II, Agadir, Morocco
- Mozambique: Data provided by / Source: Michelle Carter and Amilcar Lucas, CARE, Maputo, Mozambique
- Niger: Certifier data
- Nigeria: Data provided by / Source: Dr. Isaac Aiyelaagbe, Organic Agriculture Project in Tertiary Institutions in Nigeria, Nigeria
- Rwanda: Data provided by / Source: Rwanda Bureau of Standards/MINICO, Kigali, Rwanda
- Sao Tome and Prince: Data provided by / Source: Ecocert, Ouagadougou, Burkina Faso
- Senegal: Data provided by / Source: Ecocert, Ouagadougou, Burkina Faso

- South Africa: Data provided by / Source: Ernest Klokow, Organics South Africa, Johannesburg, South Africa; Wild collection: certifier data.
- Sudan: Only wild collection, data provided by Ecocert International.
- Tanzania: Data provided by / Source: Noel Kwai, Tanzania Organic Agriculture Movement, Dar Es Salaam, Tanzania
- Togo: Data provided by / Source: Ecocert, Ouagadougou, Burkina Faso
- Tunisia: Data provided by / Source: Prof. Ben Kheder, Centre Technique de l'Agriculture Biologique, Sousse, Tunisia
- Uganda: Data provided by / Source: Charity Namuwoza, National Organic Agricultural Movement of Uganda (NOGAMU), and Alastair Taylor, EPOPA, Kampala, Uganda
- Zambia: Data provided by / Source: Bridget O'Connor, Organic Producers & Processors Association of Zambia (OPPAZ), Lusaka, Zambia

ASIA

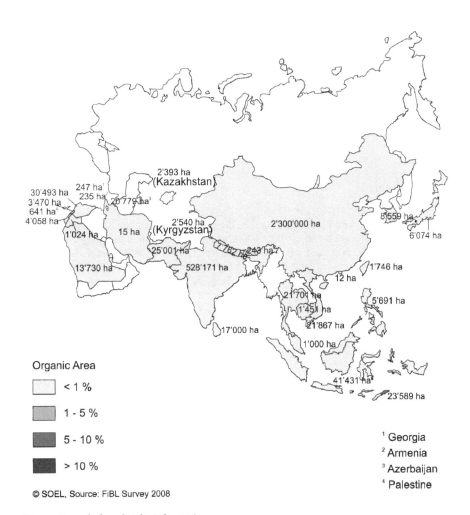

2'393 ha
(Kazakhstan)

247 ha[1]
30'493 ha
3'470 ha
641 ha[4]
4'058 ha
235 ha[2] 20'779 ha[3]
1'024 ha
15 ha
2'540 ha
(Kyrgyzstan)
25'001 ha
2'762 ha 243 ha
13'730 ha
528'171 ha
2'300'000 ha
8'559 ha
6'074 ha
1'746 ha
12 ha
21'701 ha
1'451 ha
21'867 ha
5'691 ha
17'000 ha
1'000 ha
41'431 ha
23'589 ha

Organic Area

	< 1 %
	1 - 5 %
	5 - 10 %
	> 10 %

© SOEL, Source: FiBL Survey 2008

[1] Georgia
[2] Armenia
[3] Azerbaijan
[4] Palestine

Map 3: Organic farming in Asia 2006

Organic Asia 2007

ONG KUNG WAI[1]

Organic is very much an accepted concept, and is a growing trend in Asia today; figures show continued expansion of the sector throughout the region. Although growth of the organic sector is primarily driven export markets, local markets are reportedly taking off in many of the big cities in developing countries, including in Kuala Lumpur, Manila, Bangkok, Beijing, Shanghai, Jakarta, Delhi, Bangalore and others. Japan, South Korea, Taiwan and Singapore continue to serve as engines of growth for the organic sector here, however. With a vibrant regional economy and many of the biggest cities in the world located in Asia, greater sector growth can be expected for the immediate future.

History

Pre-movement

Until the 1990s, the organic movement in the region was small. With the exception of Japan and South Korea, market activities in the cities mentioned were negligible; organic agriculture did not have much credibility in the region at the time. In fact, when activist members of IFOAM organized a pre-conference event to the IFOAM Scientific conference and general assembly in Budapest in 1990, to link up with likeminded organizations in the South, the event was billed as an 'Update on Tropical Sustainable Agriculture in Development Projects.' The term organic was intentionally not used, as the concept was not widely recognized by funding agencies at the time. Looking back, the 1990 event held at Emerson College, UK, was a benchmark meeting of development activists from Asia, Africa and Latin America, where biodynamic and organic concepts and their relevance to agriculture, rural and social development in the South were thoroughly discussed. It built bridges between the organic agriculture movements in the North, with likeminded organizations in the South. The IFOAM Third World Task Force was subsequently established by the IFOAM general assembly in Budapest, where the Southern activists also attended.

Southern activists were identified and funded through the effort of the IFOAM Third World Task Force to participate in the IFOAM Scientific conference and general assembly in Sao Paulo, in 1992. There, at a cheap downtown hotel where many of the sponsored delegates were accommodated at the time, an ad hoc evening meeting between participating Asian delegates was held. Everyone who looked Asiatic at the General Assembly was approached to attend the meeting. In the course of the meeting, the Japanese Organic Agriculture Association (JOAA) delegate made an unpremeditated offer to host an Asia organic agriculture meeting in Japan the following year. That offer marked the beginnings of the organic agriculture movement in Asia.

---

[1] Ong Kung Wai, Grolink, kungwai@tm.net.my, www.grolink.se

1st regional organic conference, Japan, 1993

The 1st Asia regional conference on organic agriculture took place in a school in Saitama prefecture outside of Tokyo during school holidays. The 20 odd delegates slept in the school dormitories and met in a classroom. It was wholly a grassroots organic community under- taking. Farmer members of JOAA delivered fresh produce. Wives and members of various Teikei groups of the association took turns to cook and serve the food at the school canteen throughout the conference. In the course of the event, development activists from develop- ing countries in the region, including Korean organic activists, got to know the Japanese organic agriculture movement and experienced Teikei, the Japanese community supported agriculture (CSA) movement.

Building on relations established in Sao Paulo, a sense of camaraderie developed from the meeting. Interest in collaborating on a regional level was established, and the South Korean delegates offered to host the next conference in Korea.

1st IFOAM Asia regional assembly, Seoul, 1995

The 2nd regional organic event, including a conference as well as an assembly of IFOAM members in the region, turned out to be a joint effort by the Korean Organic Farming Asso- ciation and the Pesticides Action Network Asia Pacific. The conference theme 'Food, Cul- ture, Trade & the Environment' represented a coming together between the organic and sustainable agriculture movements. The conference was an effort toward consensus building to address development issues related to the four subject areas of the conference theme.

Prior to the Korean conference, organic agriculture was perceived by development activists in the region as an elitist food subject with production standards set by the European and US neo-colonists. Exposure to Teikei and Community Supported Agriculture (CSA) devel- opment models allowed many to shake off their perception of organic agriculture as just a niche market phenomenon and recognize the organic movement as a credible platform for rural development and social justice. The formation of 'organic' networks and organic devel- opment agendas took place smoothly in many of the countries that now boost thriving local organic sectors. Momentum for collaboration progressed to the drafting of a regional devel- opment plan titled 'Building Alternative Systems - Community and Market development for Organic Agriculture,' at a meeting in Bangkok, Thailand, in 1996. The action plan included: standards development & harmonization; inspectors' & certification training; CSA/Teikei internships; and market development workshops. Development funding for organic was still difficult, but was becoming more accessible than earlier. However, the plan was not fully implemented due to a number of logistical and organizational challenges.

Expanding agenda and fragmentation

No organic trade actors attended the 1st regional organic conference in 1993, and few or- ganic entrepreneurs attended the 2nd conference in South Korea in 1995. A significant num- ber of Japanese organic market actors attended the 3rd IFOAM Asia conference held in Bangalore, India in 1997, and so did a host of Indian and regional social activists. The con- ference agenda expanded to a wide range of related issues, including traditional agriculture and gender in agriculture. While noted as important subjects, not as much attention was devoted towards the market, trade and the business sides of sector development as to production, extension and related social issues.

At the 1999 IFOAM Asia conference in the Philippines, a trade fair component was attached to the conference. It was small, but nevertheless showed off the local processing available in the Philippines at the time.

The 2001 conference in China had official support from the Chinese authorities, attracted good participation, and boasted the launch of a regional organic scientific research and development initiative. However, there was increasing signs that regional collaboration in development and policy advocacy was waning. In between the Philippine and China conferences, no significant regional project was undertaken by the regional board. The next regional conference scheduled for 2003 was subsequently postponed to 2004, and the regional assembly did not achieve quorum. IFOAM Asia as a regional group was consequently dissolved by the IFOAM World Board in 2005.

Growth continues apace with rising business actors and government intervention

The sector in the region has grown at an astounding rate over the last fifteen plus years, as monitored by this publication. Export opportunities have given rise to many dedicated organic enterprises and organic subsidiaries of conventional food companies. The export and trade promotion agencies of many governments in the region, including those in the Philippines, India and Thailand, have sponsored the participation of exporters at to international fairs since the late 1990s, ranging from BioFach and BioSearch Philippines to domestic fairs. In 1995, there were only a handful of market initiatives in Malaysia. Today, local organic products and imports are available in most supermarkets in major cities. Over the past decade, successful export and local niche market penetration has created a growing group of business entrepreneurs who equal - if not surpass - NGOs as the dominant sector development agents in the region, notably in Malaysia.

In general, organic sector development in the region follows two parallel models - NGO rural development project initiatives and business entrepreneurs in mostly export projects. Although marketing is acknowledged as an integral part of the organic sector, NGOs have yet to fully accept profit making as part of the organic sector culture. In order to move forward, they must resolve their suspicion of business entrepreneurs and choose to work together as co-development agents. Encouraged by market trends, many NGOs are also embarking into marketing organic products in some domestic markets under the banner of Fair Trade, with air pricing and fair returns. Supported by development funding, many forgo market margins to offer high farm gate prices. Frequently, such market initiatives are implemented without coordination with other private sector initiatives, and as a result cause further tension and precipitate calls of unfair competition by business entrepreneurs. In hindsight, opportunities to build strong mutual interests and better ties between business and development actors within the organic movement are sorely missed. This gulf is hampering the establishment of united national organic sector movement/associations in the region, which is identified as a critical factor for sector development by IFOAM in the project 'Building Sustainable Sectors' and an forthcoming publication of the same title. As Governments have emerged as proactive development actors of the organic sector in the region, this is becoming all the more important.

The need for a united private sector counterpart to facilitate effective public-private sector partnership for sector development in the region is greater than ever. There are a number of sector association initiatives: IFOAM Japan; Lanka Organic Agriculture Movement (LOAM); Organic Producers and Trade Association (OPTA), Philippines; Malaysia Organic Alliance (OAM); and the Thailand Organic Trade Association (TOTA). Most, however, have yet to gain full representative mandates. How the private sector will organize itself in the region to work with governments will determine the sector's future over the next decade.

Current developments: More of the same

In 2007, the organic sector in Asia continues robust growth in production, processing, and trade, and local market growth is on the rise as well. Building upon the success of BioFach Japan, which began seven years earlier, BioFach China was launched in Shanghai in May 2007. There were more meetings and conferences about organic sector issues, and government initiatives are becoming increasingly visible. Government officers from the region, who are responsible for developing and implementing organic policy and support programs, participated in the international Organic Agriculture Development course, sponsored by the Swedish government, marking the highest level of participation in the seven years of the course's history. Regulatory initiatives are also on the rise, according to country reports at the regional conference on organic agriculture organized by the International Trade Centre (ITC), supported by the Asia Trust Fund in Bangkok in December 2007.[1]

Growth boon and bane

Market growth in Europe and the US and its resulting growth in demand is a boon that may, however, turn into a bane for organic agriculture development in the region. Local market development in the region is in danger of being strangled by regional governments' endeavors to establish credibility and gain or protect market access to EU and US markets by establishing similar regulatory requirements, not all of which are appropriate to the production and sector development conditions in the region, including requirements for organic seeds and organic feed for livestock production.

China, which has gone on to set what is probably the most stringent of production and conformity assessment requirements, has yet to achieve any recognition from the EU, US or Japan. India, who has achieved recognition of its Program for Organic Production (NPOP) - effective only for exports - from the EU now plans to implement the same for domestic production and supply chains. The majority of local organic production in the region that is not geared to exports to the EU, US or Japan, will be subject to production requirements arising from external conditions.

Exporters have long been pushing their governments to facilitate and alleviate the cost of export certification. With good intentions, government officials are lead to believe they need to establish similar regulations to the EU and US for import recognition purposes. Unfortu-

---

[1] Information on the 'Regional Conference on Organic Agriculture in Asia,' held December 15-17, 2007 in Bangkok, Thailand financed by the European Commission and the International Trade Centre (ITC) under the Asia Trust Fund is available at the ITC homepage at http://www.intracen.org/organics/asia_conference.htm

nately, such export-inspired requirements are likely to be prematurely applied on to the domestic sector, crippling a nascent domestic market sector with undue stringent production and conformity assessment requirements. These are requirements the EU and US organic industry themselves did not have to contend with at a similar early stage of their own development.

Future growth lies within

According to UNCTAD, up to 70 percent of trade in agriculture products in Asia is inter-regional. Although inter-regional trade in organic products is insignificant today, it can be expected to match conventional trade flows as the sector becomes more mainstream. Whether or not regulators have factored this longer term scenario in their norm setting initiatives is not clear. Nevertheless, it is time - sooner than later - to review regulatory initiatives in the region to avoid them becoming trade obstacles in the future.

Coming at the heels of the ITF[1] meeting in Bali (November 26-29, 2007), where the 1st draft of a Tool for Equivalence of Organic Standards and Technical Regulations (EquiTool) was discussed and the 4th draft of the International Requirements for Organic Certification Bodies (IROCB) was reviewed, the speakers from the United Nations Conference on Trade and Development (UNCTAD) and IFOAM at the ITC conference in Bangkok[2] called for a increase in regional collaboration in norm setting and certification. The East African example, where agreement on a common set of organic standards and mark was reached between Kenya, Uganda and Tanzania, was presented as food for thought[3].

The calls also echo the findings and outcomes of the conference of the Asia-Pacific Economic Cooperation (APEC)[4] held earlier in October 2007 in Beijing. Identifying duplication of certifications for different target markets as an obstacle to trade, the importance and necessity of cooperation and recognition within APEC for the benefit of consumers, operators, certification bodies and governments was highlighted. A working group to coordinate cooperation and recognition of organic certification among APEC members was proposed to the APEC Secretariat.

For more information on standards, regulations and certification, see the chapters by Herrmann, Rundgren, Huber et al. and Twarog in this book.

Regional collaboration anyone?

Will regional collaboration in Asia re-emerge? Collaboration in the region is evidently alive and well; a group of certification bodies and initiatives met for the second time and finalized a draft Memorandum of Understanding (MoU) for collaboration in inspection and certifica-

---

[1] International Task Force on Harmonization and Equivalence in Organic Agriculture (ITF)
www.unctad.org/trade_env/itf-organic/welcome1.asp
[2] Information on the 'Regional Conference on Organic Agriculture in Asia,' held December 15-17, 2007 in Bnkok,
Thailand financed by the European Commission and the International Trade Centre (ITC) under the Asia Trust Fund is
available at the ITC homepage at www.intracen.org/organics/asia_conference.htm
[3] See also chapter by Sophia Twarog on the East African Organic Standard in this book.
[4] Homepage of the Asia-Pacific Economic Cooperation (APEC): www.apec.org/

tion, in Vientiane, Laos, in November 2007. Participants reportedly included delegates from China, Laos, Indonesia, Italy, Malaysia, Nepal, New Zealand, Philippines, South Korea, Sri Lanka Thailand and Vietnam. The MoU outlines a collaboration framework and procedures to offer a regional one stop service for organic producers seeking international certification for organic products in the region

At another development planning meeting in the region, in late November 2007 in Kuala Lumpur, the IFOAM Development Options Task Force proposed to offer sector development options workshops to assist government officials and private sector development actors in their sector building efforts. One workshop is tentatively scheduled to take place in Asia in 2008, with the intention of facilitating consensus building and uniting the organic private sector to facilitate effective public-private sector partnerships for sector development in the region. The future is ours to lose.

Country Updates

For reports on China, Thailand, Vietnam and Indonesia, see the 2007 edition of 'The World of Organic Agriculture' (Ong 2007).

Japan

The Japanese market received a boost from the upswing in ecological awareness arising from the implementation of the Kyoto Protocol and the introduction of the LOHAS (*Lifestyles of Health and Sustainability*) concept by the lifestyle magazine, Sotokoto. Organic became fashionable along with Eco lifestyle and Slow Food, receiving more media coverage in 2006 than ever before, according to a local industry observer. This trend has attracted a new and younger generation of people to the organic community. There is an explosion of organic related information posted through personal web logs and numerous internet exchange communities.

Unlike China, organic production in Japan, measured at 0.2 percent of total domestic agricultural production, is at a standstill. This is nevertheless good news according to the Japanese movement. Many had projected and feared a big drop from competing imports. The increase in sales has mainly been of processed foods using imported ingredients. Foreign production of JAS certified produce is far higher than domestic production (see table). Most foreign country production is processed when entering the Japanese market. Local producers face competing imports even in traditional foods such as Konyaku jelly.

The revised JAS regulation was implemented in March 2006; after that, all certified organic operators needed to re-certify under revised JAS regulation. The number of certified organic farms before the revision was 3'592, and after the revision 2'258, mainly due to the double counting of many farms. One reason for the slow and low rate of renewals of certified domestic producers, according to the Japanese movement, is that certification does not really help to boost sales. Farmers' markets nevertheless remain popular.

Table 16: Domestic production and imports to Japan 2006

Products	Domestic production (tons)	Imports (tons)
Vegetables	29'949	106'119
Fruits	1766	131'538
Rice	10'811	21'777
Wheat	558	7528
Soybeans	974	63'647
Beans	110	21403
Miscellaneous	26	5085
Green tea	1538	449
Black tea	2	639
Coffee bean	0	6070
Nuts	0	9578
Sugar beets	0	83'9937
Arum root	1365	3743
Palm fruits	0	68'873
Other	1496	8880
Total	48'596	1295'266

Source: Data provided by IFOAM Japan

The parliamentarians' advocacy group for organic agriculture established in 2004 managed to push through an organic agriculture promotion bill in 2006. IFOAM Japan expects that organic agriculture will grow. Currently, the local governments are preparing guidelines to promote and support organic agriculture.

Republic of Korea

Currently, there are more than 8'000 hectares under organic management in Korea; most of the certified organic farmers are vegetable producers, growing up to 30 different vegetables. In the early 1990s, the government decided to adopt a sustainable agriculture development policy. The direct payment program was adopted in 1999. No organic agricultural products and foods are exported to foreign countries, and no foreign certification body is operating in Korea. A lot of processed foods (baby food, juice, wine, ice cream, pasta, olive oil, cereals, cheese, chocolate, cloth, cosmetics) are imported from Germany, the UK, France, the US, Australia and New Zealand. Fresh organic products are mainly imported from China, the US, Australia and New Zealand.

Imports of non-processed organic products were 6'843 tons in 2006, according to the Ministry of Agriculture MAF (904 tons in 2003). Most of the organic grains are imported from China and are being used to produce processed foods, such as tofu. However, fresh organic fruits and vegetables are not commonly imported due to the restrictive government regulations and transportation problems of perishable products.

In 2005, one of the largest organic foods distributors in Korea imported 200 new products: 50 percent from the US, 40 percent from the European Union and ten percent from Japan.

The Korean organic foods market, even though still small, has shown strong growth over the past five years, and local industry sources project a growth rate of 30 to 40 percent annually. Organic products can increasingly be found in mainstream retail outlets.

The Korean government has developed an organic certification and labeling program that has helped to increase consumer awareness. At present, one US organic farm has been approved by a Korean certification bodies to export organic vegetable seeds (broccoli, alfalfa and radish) to Korea.

The regulations for fresh organic produce and grains are implemented by the Ministry of Agriculture and Forestry (MAF), and the regulations for processed organic products are implemented by the Korean Food and Drug Administration (KFDA). The Research Committee of Organic Food in KFDA has worked for last two years to finalize the new regulation on organic food.

Furthermore, the Korean government encourages farmers to grow organic products, and participating farmers receive a premium for their products. The government also provides certain subsidies to advertise organic agricultural products and for packing organic products.

Since Spring 2005, the Research Institute of Organic Agriculture of Dankook University has offered courses on organic agriculture teaching the principles of organic agriculture and practical skills of organic rice, fruit and vegetable cultivation, and organic animal husbandry. Dankook University also offers Master and PhD courses on organic agriculture.

India

Organic farming is developing very dynamically in India. The organic land area has increased substantially between 2005 and 2006 and is now more than 500'000 hectares. The Indian authorities managed to acquire both USDA equivalence for the NOP and the EU third country listing in 2006. Furthermore, recognizing the difficulty smallholders face to access third party certification, the government launched a national participatory guarantee system program, with the support of the FAO India office to facilitate organic assurance.

According to the Press Information Bureau Government of India, the Government is implementing a National Project on Organic Farming (NPOF) for production, promotion, certification and market development of organic farming in the country. Financial assistance is being provided for the capacity building through service providers, setting up of organic input production units, promotion of organic farming through training programs, field demonstrations, setting up of model organic farms and market development.[1]

The Indian organic produce is mainly exported to Japan, Netherlands, Italy, France, Switzerland, the UK and the US. According to the Indian Centre for Organic Agriculture (ICCOA), a major reason for the growth in organic farming is increased awareness among consumers in the country, even though until recently food was mainly being exported. But over the last couple of years, the domestic market has started growing. The third edition of India's first organic products trade fair, India Organic 2007, saw participation of a record 184 companies

---

[1] Government of India, Press release of November 19, 2007 at http://pib.nic.in/release/release.asp?relid=32798

and 25 state governments and 12 countries. The fair witnessed 15'000 business visitors and key policy makers from the Central ministries and state governments (Kulkarni 2008).

Table 17: Major organic products exported from India 2006/2007

Products	Export Quantity (in metric tons)	Destination
Honey	1'783	Germany, US,. UK
Spices	742	Italy, US, Sri Lanka, UK South Africa
Basmati Rice	5139	US, Kuwait, Netherlands, Germany, Mauritius, UK, Israel, New Zealand, Sweden
Cereals (including non basmati Rice)	97	US, Germany
Tea	3'026	Australia, Czech. Rep., France, Germany, UK, US, China, Canada, Taiwan, Switzerland
Coffee	141	Belgium, Germany, US.
Medicinal & Herbal Plants	426	US, Taiwan, Malaysia
Others	8'098/	US, Switzerland, UAE, EU, Pakistan, (cotton), Indonesia (cotton), Japan, Denmark (walnut)
Total	19'456	

Source: APEDA 2007

Acknowledgements

The following persons provided input the country reports: Katsu Murayama, Yoko Taniguchi, Satoko Miyoshi (Japan), Sang Mok Sohn (Republic of Korea), Manoj Menon Kumar (India).

References and Links

IFOAM (2008): Examples and recommendations for development options for the emerging organic sector, based on analytical surveys of cases around the world. International Federation of Organic Agriculture Movements (IFOAM), Bonn

IFAD (2005): Organic Agriculture and Poverty Reduction in Asia: China and India Focus. Document of the International Fund for Agricultural Development, Report no. 1664, July 2005.

International Trade Centre ITC: Country Profiles Asia. The ITC Trade Centre. The ITC homepage.
    http://www.intracen.org/organics/Country-Profiles_overview.htm

Kledal, Paul Rye , Qiao Yu Hui, Henrik Egelyng, Xi Yunguan, Niels Halberg and Li Xianjun, Country report: Organic food and farming in China. IFOAM/FiBL (2007): The World of Organic Agriculture. Statistics and Emerging Trends 2007. IFOAM, Bonn & FiBL, Frick

Ong, Kung Wai (2007) Organic Agriculture in Asia. In: IFOAM/FiBL (2007): The World of Organic Agriculture. Statistics and Emerging Trends 2007. IFOAM, Bonn & FiBL, Frick

Thode Jacobsen, Birthe (2007): The European Market for Organic Fruit and Vegetables from Thailand' . This report was undertaken on behalf of the International Trade Centre UNCTAD/WTO (ITC) and was funded under the EU-ITC Project Strengthening the Export Capacity of Thailand's Organic Agriculture (Asia Trust Fund). ITC, Geneva http://www.intracen.org/Organics/documents/Thailand-fruit-and-vegetables.pdf

USDA, 2006: China, Peoples Republic of Organic products and Agriculture in China 2006. USDA Foreign Agricultural Service, GAIN report number CH6405, 21 of June 2006. Report prepared by Yaang Mei, Michael Jewison, Christina Greene. www.fas.usda.gov/gainfiles/200606/146198045.pdf

Country Report: Organic Agriculture in Iran

HOSSEIN MAHMOUDI AND ABDOLMAJID MAHDAVI DAMGHANI[1]

Introduction

Iran occupies a vast land area with diverse climatic conditions, and as a result exhibits rich biological diversity. Evidence indicates that the country has served as a center of agricultural and human evolution for at least the last 100'000 years. Traditional small scale farming constituted the makeup of farming communities for centuries, resulting in the tremendous accumulation of indigenous knowledge in farming practices and food production (Koocheki, 2004).

About 86 percent of farmers in Iran are smallholders; they manage nearly 40 percent of the arable land in Iran, which is mainly used for producing high-value cash crops such as rice, saffron and vegetables (Mahmoudi et al., 2007). Smallholders cultivate much of the land without resorting to agrochemicals, and traditional mixed farming systems remain prevalent. In small farming systems, which account for more than 80 percent of agricultural production in Iran, ecological practices are prevalent.

Organic agriculture offers the potential to enable Iranian smallholders to achieve household food security and increase income, while regenerating the land, enhancing biodiversity, and supplying quality food to local communities. Correspondingly, there is an increasing public concern about food safety, but comparatively few people in Iran are aware of organic farming. A recent survey indicated that there is lack of information on organic farming. (Ghorbani et al, 2007). The organic sector still has a long way to go in Iran.

Organically cultivated area in Iran

At this time, producers are not paying for the costs of certification; export companies manage arrangements for inspections by foreign certification bodies and bear the resultant costs. The number of organic initiatives in Iran are few; an 11-hectare saffron farm is certified by a French company, an 85-hectare olive orchard that sells its products in local markets, and there are 22'000 hectares of fig production. Although none of them are certified organic according to national standards or regulations, a fraction of the above-mentioned fig production has been certified according to the AB standards (Agriculture Biologique, France), and the products are exported to Europe.

Standards, certification and regulation

No official legislation for organic agriculture has been passed in Iran. However, the Shahid Beheshti University, in cooperation with the Institute of Standards and Industrial Research of Iran, has taken the initial actions towards the development of a national standard for organic production and processing. The draft of this standard was released in November 2007.

---

[1] Environmental Sciences Research Institute, Shahid Beheshti University, Tehran, Iran

Government policies regarding organic agriculture

The government has started a program to reduce the use of agrochemicals. Governmental subsidy of agrochemicals has been reduced dramatically during the past year, which is a step in the right direction for the development of organic agriculture. Furthermore, the 'Committee on Organic Agriculture' was recently established in the Ministry of Agriculture in order to develop organic farming policies and provide an action plan for development of organic agriculture in Iran.

Research and education

Recently, a research program on organic agricultural production, processing and marketing was started by institutes the Environmental Sciences Research Institute at the Shahid Beheshti University. Also, a postgraduate course on agroecology started at the Shahid Beheshti University and at the Ferdowsi University of Mashad.

References

Ghorbani, Mohammad; Mahmoudi, Hossein and Liaghati, Houman (2007) Consumers' Demands and Preferences for Organic Foods: A Survey Study in Mashhad, Iran. Poster presented at the 3rd QLIF Congress: Improving Sustainability in Organic and Low Input Food Production Systems, University of Hohenheim, Germany, March 20-23, 2007. http://orgprints.org/9831/

Koocheki, A. (2004). Organic farming in Iran. 6th IFOAM-Asia Scientific Conference 'Benign Environment and Safe Food', 7th – 11th September 2004. Yangpyung/ Korea.

Mahmoudi, H., H. Liaghati and Majid Zohari (2007). The Role of Organic Agriculture in Achieving the Millennium Development Goals: challenge and Prospects in Iran. Conference of Tropentag 2007: Utilisation of diversity in land use systems: Sustainable and organic approaches to meet human needs. October 9 - 11, 2007, Witzenhausen, Germany

Asia: Organic Farming Statistics

Table 18: Organic agricultural land and farms in Asia 2006

Country	Organic agricultural area (ha)	Share of agricultural land	Farms
Armenia	235	0.02%	35
Azerbaijan	20'779	0.44%	388
Bhutan	243	0.04%	53
Cambodia	1'451	0.03%	3'628
China	2'300'000	0.41%	1'600
Georgia	247	0.01%	47
Hong Kong (2005)	12	No data	20
India	528'171	0.29%	44'926
Indonesia	41'431	0.09%	23'608
Iran	15	0.00%	2
Israel	4'058	0.71%	216
Japan	6'074	0.16%	2'258
Jordan	1'024	0.09%	25
Kazakhstan	2'393	0.00%	No data
Korea, Republic of	8'559	0.45%	7'167
Kyrgyzstan	2'540	0.02%	392
Lebanon	3'470	1.02%	213
Malaysia	1'000	0.01%	50
Nepal	7'762	0.18%	1'183
Pakistan	25'001	0.10%	28 (2004)
Palestine	641	0.19%	303
Philippines	5'691	0.05%	No data
Saudi Arabia (2005)	13'730	0.01%	3
Sri Lanka	17'000	0.72%	4'216
Syria	30'493	0.22%	3'256
Taiwan (March 2007)	1'746	0.21%	905
Thailand	21'701	0.12%	2'498
Timor Leste	23'589	6.94%	No data
Vietnam	21'867	0.23%	No data
Total	3'090'924	0.17%	97'020

Source: FiBL Survey 2008

Table 19: Use of organic agricultural land in Asia 2006

Main land use type	Crop category	Agricultural area (ha)
Arable land	Arable crops, no details	14'584
	Cereals	33'320
	Fallow land as part of crop rotation	306
	Flowers and ornamental plants	36
	Green fodder from arable land	265
	Industrial crops	1'576
	Medicinal & aromatic plants	6'992
	Oilseeds	3'683
	Protein crops	9
	Root crops	271
	Seeds and seedlings	3
	Textile fibers	29'051
	Vegetables	3'777
Arable land		*93'873*
Permanent crops	Citrus fruit	864
	Coffee	50'184
	Fruit and nuts	7'185
	Grapes	481
	Medicinal & aromatic plants	182
	Olives	1'437
	Other permanent crops	201
	Tea	3'358
	Tropical fruit and nuts	2'234
Permanent crop, total		*66'126*
Cropland, other/no details		*998'123*
Permanent grassland		*711'452*
Other		*1'034*
No information		*1'220'315*
Total		3'090'924

Source: FiBL Survey 2008

Table 20: Certified wild collection in Asia 2006

Country	Crops harvested	Organic wild collection area (ha)
Armenia	Fruit and berries	83'000
Azerbaijan	Fruit and berries	313
	Nuts	175
	Wild collection, no details	3'200'000
	Total Azerbaijan	3'200'488
China	Wild collection, no details	1'900'000
India	Wild collection, no details	2'432'500
Indonesia	Bee pastures	7'378
Kyrgyzstan	Wild collection, no details	55'086
Lebanon	Wild collection, no details	14
Nepal	Wild collection, no details	24'421

Country	Crops harvested	Organic wild collection area (ha)
Vietnam	Wild collection, no details	1'368
Mongolia	Wild collection, no details	360'000
Total		8'064'255

Source: FiBL Survey 2008

Data and information sources/contacts

- Armenia: Data provided by / Source: Nune Darbinyan, Ecoglobe, Yerevan, Armenia
- Azerbaijan: Data provided by / Source: Vugar Babajev, Ganja Agribusiness Bhutan, Ganja, Azerbaijan
- Bhutan: Data provided by / Source: Kesang Tshomo, Biobhutan, Thimpu, Buthan
- Cambodia: Winfried Scheewe, DED Advisor to CEDAC, Phnom Penh, Cambodia
- China: Data provided by / Source: Dr. Wang Maohua, Department for Registration, Certification and Accreditation Administration (CNCA), Beijing, China. Land use according to Paul Kledal, University of Copenhagen, Copenhagen, Denmark
- Georgia: Data provided by / Source: Mariam Jorjadze, Biological Farming Association Elkana, Tbilisi, Georgia
- India: Data provided by Manoj Kumar Menon, ICCOA, Bangalore (for total organic area and farms) and by Dr. Gouri, Agricultural and Processed Food Products Export Development (APEDA) (for production data). Source: APEDA, Ministry of Commerce, Govt of India, New Delhi, India and National Centre for Organic Agriculture (NCOF)
- Indonesia: Data provided by Agung Prawoto, BIOCert, Bogor, Indonesia, www.biocert.or.id, based on certifier data
- Iran: Data provided by / Source: Hossein Mahmoudi, Environmental Science Research Institute, Shahid Beheshti University, Tehran, Iran
- Israel: Data provided by / Data source: Pnina Oren Shnidor, Plant Protection and Inspection Services
- Japan: Data provided by Satoko Miyoshi, IFOAM Japan, Toda city, Saitama, Japan/ Source: Ministry of Agriculture MAFF. Calculation of organic land by multiplication of the number of farms with the average size of Japanese farms.
- Jordan: Data provided by Lina Al Bitar, IAMB Bari; Source: Mediterranean Organic Agriculture Network (MOAN); Bari, Itay
- Kazakhstan: Certifier data
- Korea, Republic of: Data provided by / Source: Prof. Dr. Sang Mok Sohn, Research Institute of Organic Agriculture, Dankook University, Cheonan, Republic of Korea
- Kyrgyzstan: Source: Certifier data. Cotton data provided by Helvetas, Zurich, Switzerland
- Lebanon: Data provided by / Source: Nada Omeira, Association for Lebanese Organic Agriculture (ALOA), Beirut, Lebanon
- Malaysia: Data provided by / Source: Ong Kung Wai, Grolink, Penang, Malaysia
- Mongolia: Certifier data for wild collection
- Nepal: Data provided by / Source: Maheswar Ghimire, Katmandu, Nepal
- Pakistan: Certifier data for total organic land (2006); land use data as of 2004; provided by Ghulam Mustafa, Pakistan Organic Farms (POF), Lahore, Pakistan
- Palestine: Data provided by Lina Albitar, IAMB Bari; Source: Mediterranean Organic Agriculture Network (MOAN); Bari, Itay
- Philippines: Data provided by Lani Limkin, Organic Certification Center of the Philippines (OCCP)

- Saudi Arabia: Data provided by / Source: Shafiq Urrahman, Watania Agriculture Establishment, Riyadh, Saudi Arabia
- Sri Lanka: Data provided by / Source: Nalika Kodikara, Sri Lanka Export Development, Colombo and Saminathan Vaheesan, Helvetas Sri Lanka, Colombo
- Syria: Land use data provided by Dr. Souhel Makhoul, General Commission for Scientific Agricultural Research (GCSAR). Syria; Totals according to Mediterranean Organic Network, c/o IAMB Bari, Italy
- Taiwan: Data provided by / Source: Perrine Liu, Yu-shi Co., Taipei, Taiwan
- Thailand: Data provided by / Source: Vitoon Panyakul, Green Net, Bangkok
- Timor Leste: Certifier data
- Vietnam: Certifier data

EUROPE

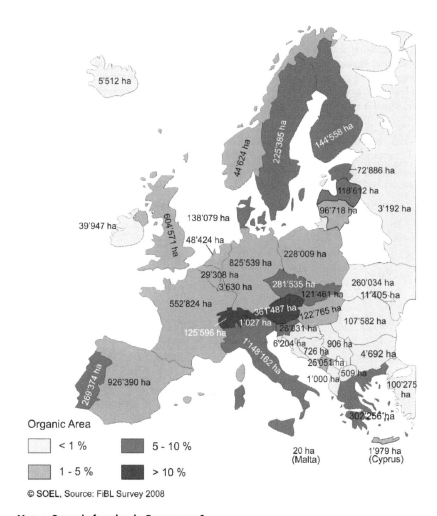

Organic Area

<table>
<tr><td>☐</td><td>< 1 %</td><td>■</td><td>5 - 10 %</td></tr>
<tr><td>☐</td><td>1 - 5 %</td><td>■</td><td>> 10 %</td></tr>
</table>

© SOEL, Source: FiBL Survey 2008

Map 4: Organic farming in Europe 2006

117

Europe: Statistics, Policy and Research

HELGA WILLER[1]

Statistical Development: Continued Growth

Since the beginning of the 1990s, organic farming has rapidly developed in almost all European countries. Almost 7.4 million hectares (1.6 percent of the agricultural land) were managed organically by more than 200'000 farms in 2006. In the European Union (EU-27), almost 180'000 farms managed 6.8 million hectares organically, constituting four percent of the agricultural area.

Compared to 2005, the organic land increased by 526'562 hectares (+7.7 percent), due to substantial increases in Spain (+118'821 hectares), Italy (+81'060 hectares), Poland (+68'300 hectares), Portugal (+56'646 hectares) and other countries. There have been substantial relative increases in many Eastern and South Eastern European countries; in Croatia and Macedonia, for instance, the organic area has doubled.

Development of the organic agricultural land in Europe 1985-2006

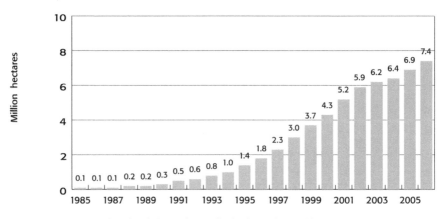

Source: Institute of Rural Sciences, Aberystwyth University, UK; FiBL, CH-Frick

Figure 17: Development of organic land area in Europe 1985-2006

Source: Institute of Rural Sciences, Aberystwyth University, UK and FiBL Frick, Switzerland

---

[1] Helga Willer, Research Institute of Organic Agriculture (FiBL), Ackerstrasse, 5070 Frick, Internet www.fibl.org

It is expected that the organic land area also increased in 2007. Austria and the Czech Republic were the first countries to communicate new data. In the Czech Republic, the area was 312'890 hectares, an increase of more than 30'000 hectares, and Austria reported a 10'000 hectare increase with 371'000 hectares. (These data are not in the table at the end of the chapter, which presents the situation as of 2006).

The difference between individual countries regarding the importance of organic farming is substantial. More than 13 percent of agricultural land in Austria is organic (2007 estimates at 16 percent), 12 percent in Switzerland, and 9 percent in Italy and Estonia. The country with the highest number of farms and the largest organic land area is Italy, followed by Spain, Germany and the UK.

The ten countries with the largest areas of organic agricultural land in Europe 2006

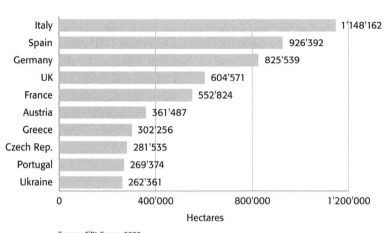

Source: FiBL Survey 2008

Figure 18: The ten countries with the largest area of organic land in Europe 2006

Source: FiBL survey 2008

In Europe, the organic agricultural land (7.4 million hectares) is mainly used for permanent pastures (44 percent) and for arable cropping (41 percent). Permanent crops account for 9 percent of the land. Information about the main land use categories was available for 99 percent of the organic land. Cereals and fodder crops play the most important role in arable farming. Among the permanent crops, olives, fruits, nuts, and grapes are the most important categories; for the European Union, a similar picture emerges (see following article by Diana Schaack). Furthermore, Europe has more than 9.5 million hectares of certified organic wild collection. Most of this land is in Finland, followed by Serbia and Bosnia Herzegovina.

Policy

Support for organic farming in the European Union includes support under the European Union's rural development programs,[1] legal protection under the recently revised EU regulation on organic faming[2] and the launch of the European Action Plan on Organic Food and Farming in June 2004.Sources:[3]

The **Rural Development policy 2007-2013**,[4] as set forth in Council Regulation (EC) No 1698/2005 of 20 September 2005,[5] employs targeted measures for strengthening rural development instead of a general support to the agricultural sector. Under the rural development programs, organic farming has been supported with area payments since 1992, and these payments are one reason for the high number of organic farms in the European Union and other European countries, many of which have similar schemes. Under the rural development policy 2007 to 2013, organic farming is supported in most countries by area based payments to farmers; these payments differ, however from country to country or even within one country. In Germany, for instance, payments for arable land are between 120 and 190 Euros per hectare and year, and some federal states grant up to 262 Euros during the conversion period. The rates for permanent grassland range between 120 and 187 Euros, and for permanent crops - like grapes and fruit - between 380 and 1020 Euros per hectare and year.[6] The draft Rural Development Program of the Slovak Republic for 2007 – 2013 allocates financial support for organic farming ranging from approximately 130 Euros for permanent grass cover to approximately 903 Euros for orchards and vineyards. The financial support for arable land is approximately 205 Euros and for vegetables, and for medicinal plants, spice and aromatic plants, the payment is approximately 665 Euros per hectare (Source: Ministry of Agriculture of the Slovak Republic according to Lehocka & Klimekova 2008[7]).

At its European Organic Congress in December 2007, the IFOAM EU Group concluded it should be insured that some of the proposed increase in rural development funding is targeted at organic farming, in particular in regions and member states that have so far given a low priority to organic farming support.[8]

---

[1] A collection of links related to European agripolicy documents is available at www.organic-europe.net/europe_eu/rural-development.asp

[2] Council Regulation (EEC) No 2092/91 of 24 June 1991 on organic production of agricultural products and indications referring thereto on agricultural products and foodstuffs; available via www.organic-europe.net/europe_eu/eu-regulation-2092-91.asp

[3] Information on the European Action plan is available at
http://europa.eu.int/comm/agriculture/qual/organic/plan/index_en.htm.

[4] European Commission (2006): Rural Development policy 2007-2013. The European Commission's Homepage, Brussels, http://ec.europa.eu/agriculture/rurdev/index_en.htm

[5] COMMISSION REGULATION (EC) No 1320/2006 of 5 September 2006 laying down rules for the transition to the rural development support provided for in Council Regulation (EC) No 1698/2005

[6] A detailed list of the rates in all German federal states is available at the Central Internet Portal for Organic Farming www.oekolandbau.de at www.oekolandbau.de/erzeuger/umstellung/foerdermittel/

[7] Zuzana Lehocká and Marta Klimeková (2008): Organic Farming in the Slovak Republic 2007. The Organic Europe Homepage. FiBL, CH-Frick. Available at http://www.organic-europe.net/country_reports/slovakia/default.asp; January 31, 2008

[8] IFOAM EU Group, Press release of December 5, 2007: Organic congress concludes with new political agenda. Brussels. Available at www.ifoam.org/about_ifoam/around_world/eu_group/PR_OrganicCongress_5.12.2007.pdf

Regulation: On July 20, 2007 the new organic regulation was published - Council Regulation (EC) No 834/2007 of 28 June 2007 on organic production and labelling of organic products and repealing Regulation (EEC) No 2092/91.[1] It will come into force on January 1, 2009. According to a press release of the European Commission,[2] the new rules set out a complete set of objectives, principles and basic rules for organic production, and include a new permanent import regime and a more consistent control regime. The use of the EU organic logo will be mandatory, but it can be accompanied by national or private logos. The implementation rules are expected to be published in mid of 2008. Most likely, many European States that are not members of the EU will start to adapt their regulations to the EU Regulation in 2008/2009.

The information campaign proposed in the **European Action Plan for Organic Food and Farming** is expected to start in 2008. With this campaign, Action 1 - a multi-annual EU-wide information and promotion campaign to inform consumers, public institutions canteens, schools and other key actors - will be implemented. This campaign is funded under Council Regulation (EC) No 2826/2000 of 19 December 2000 on information and promotion actions for agricultural products on the internal market.[3] With such funds, several national information campaigns have already been or are currently being co-funded, for instance the German campaign 'Bio – mir zuliebe' or the French 'Printemps Bio.'

Research

Today, organic farming research is substantially funded under national research programs or national organic action plans, as well as through European projects.[4] A unique overview of research currently conducted in Europe was given at the Joint Organic Congress, which was held in May 2006 in Odense, Denmark, where 300 papers were presented.[5] Further results were made available at the 3rd QLIF Congress, held March 20-23, 2007 in Germany.

Even though no figures for all European countries are available, it is known that the funds of the eleven countries that are part of the ERA-Net project CORE Organic (see below) and of the European Union amounted to approximately 64 million Euros for organic farming research in 2006.

The establishment of joint funding for transnational research in organic food and farming

---

[1] The regulation is available at http://eur-lex.europa.eu/LexUriServ/site/en/oj/2007/l_189/l_18920070720en00010023.pdf

[2] European Commission, press release of June 12, 2007: Organic Food: New Regulation to foster the further development of Europe's organic food sector. Available at http://europa.eu/rapid/pressReleasesAction.do?reference=IP/07/807&format=HTML&aged=0&language=EN&guiLanguage=en

[3] European Commission: Promotion for EU agricultural products on the Internal Market. The European Commission Homepage, http://ec.europa.eu/agriculture/prom/intern/index_en.htm

[4] For a list of projects funded by the European Commission see www.organic-europe.net/europe_eu/research-euprojects.asp

[5] The papers can be downloaded at the Organic Eprints Archive at http://orgprints.org/view/projects/int_conf_joint2006.html. Organic Eprints is an internet-based archive for papers related to research in organic agriculture. The database has now more than 5000 entries.

was one of the main objectives for the ERA-NET Project CORE Organic.[1] The eleven CORE member countries have contributed with 100'000 to 600'000 Euros each per year to this pilot program, which will run from 2007–2009. Under the CORE Organic pilot call for transnational research in organic food and farming, eight projects that began in 2007 have received funds.

The CORE Organic Project at www.coreportal.org portal has country reports on organic food and farming research in the eleven CORE Organic partner countries. A summary of these reports is below (see also Lange et al. 2006).

Funding for organic food and farming research in 11 European countries 2006

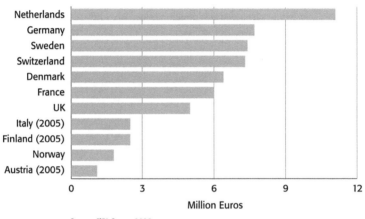

Source: FiBL Survey 2006

Figure 19: Funding for organic food and farming research in 11 European countries 2006

Source: Country reports in Lange et al., 2006; also available at www.coreportal.org, with updates of the CORE Organic partners

- In **Austria**, organic farming research started in 1980, initiated by the private Ludwig-Boltzmann Institute for Organic Farming and Applied Ecology in Vienna (now Bio Forschung Austria). In 1996, the Institute for Organic Farming at the Agricultural University in Vienna (BOKU) was set up in order to intensify teaching of students and to do research. Most recently, the national Centre for Agricultural Research at Raumberg Gumpenstein set apart an institute dedicated to organic farming research. In addition, the Veterinary University in Vienna and the University of Innsbruck are involved in organic farming research. In Austria, organic farming is one topic of the national research program PFEIL 05 (2002 to 2005), which was funded by the Federal Ministry of Agriculture, Forestry, Environment and

---

[1] CORE Organic (Co-ordination of European Transnational Research in Organic Food and Farming); Internet www.coreorganic.org. CORE Organic is a three year Co-ordination Action in organic food and farming (2004 to 2007).

Water Management. The new program (PFEIL 10, from 2006 to 2010) also funds organic farming research. A network of scientists and stakeholders called 'BioEnquête' helped to set priorities and to survey the research activities.

- **France** has several government programmes for organic farming research; most of this research is carried out at state research institutes like the National Institute of Agronomic Research INRA. Furthermore research activities carried out at private institutes like the Technical Institute for Organic Farming ITAB and the Research Group on Organic Farming GRAB. France spends about 6 million Euros on organic farming research annually. Between 2000 and 2006, almost 40 million Euros were spent.

- **Germany** was a pioneer country in organic farming research. The biodynamic research institute in Darmstadt was already funded in 1950, the first university chair dedicated only to organic food and farming started 1981 at the Kassel University in Witzenhausen, the second University chair 1987 in Bonn, followed by others since. In 1996, the entire Faculty for Agronomy at Kassel University with 20 chairs got oriented towards organic food and farming research. In 2000, the Federal Agricultural Research Centre (FAL) in Braunschweig established an institute for organic farming (in the very Northern part of Germany). In 2001, the German Ministry established - after a wide consultation with all stakeholders - the Federal scheme on organic farming that funded research activities in 2002 and 2003 with 10 million Euros annually. For the second phase from 2004 to 2007, 7 million Euros on average were allocated to research projects every each year. The program will be continued until 2009 at least. The communication of research activities is through the internet site forschung.oekolandbau.de. Research results are available at the open access archive Organic Eprints (www.orgprints.org/).

- **Italy:** In 2001, the Italian Ministry for Agriculture MIPAF launched an action plan on organic farming research. This action plan was worked out after consultations with the regions and with the national committee for organic agriculture. The first open call was published in 2002 and many research projects are still ongoing. A broader action plan focusing on research and on promotion of organic farming was launched for the period from 2005 to 2007. The main actors are the National Centres for Research (CRA) and various agricultural universities. Many universities offer Master Courses in organic farming, and the Research Centre for Mediterranean Agriculture in Bari (IAMB-CIHEAM) is an international player in organic farming training and research.

- In the **Netherlands**, the Ministry of Agriculture, Environment and Food Quality introduced a policy in favor of organic farming with the goal to convert ten percent of the agricultural land area by 2010. In parallel, the Ministry increased the funding for organic farming research from 4.6 million Euros in the year 2000 to 11.2 million in the year 2006; totaling at 66 million Euros. In order to integrate the stakeholders into priority setting and into the annual work program, the stakeholder network 'Bioconnect' funded by Ministry is in charge of consultation and supervising work. The funding goes to the Wageningen University and Research Center (WUR). There is a bilateral contract between WUR and the private Louis Bolk Institute (an early pioneer institute in organic farming research) on many research projects, so that part of the funding goes to Louis Bolk Institute as well. Commercial farmers are closely linked to the research projects. Therefore, results from research work have been very relevant to the practice, and the mutual flow of information is strong.

- In **Norway**, research activities were started in 1986 by the private Institute NORSØK at Tingvoll. Twenty years later, in 2006, NORSØK was merged with two state research institutes now called Bioforsk, and organic farming is a small part of their activities. In addition to Bioforsk, several other universities and research centers are involved in organic farming research.

- In **Sweden**, the first national research program on organic farming was launched in 1996, funded by the Forestry and Agriculture Research Board (SJFR), now Swedish Research Council for Environment, Agricultural Sciences and Spatial Planning (Formas). Since 1996, there has been a permanent sequence of three-year programs on organic farming. The total funding for organic farming research was almost 50 million Euros 2000 to 2006. The Center for Sustainable Agriculture (CUL) at the Swedish University of Agricultural Sciences (SLU) coordinates the programs 'Formas' and 'Ekoforsk.' The coordination is accompanied by a group of stakeholders. The research projects are mainly run by the run by the Universities of Halmstadt, Linköping, Lund, Uppsala, by SLU (main actor), and by the two private Institutes.

- **Switzerland** was an important pioneer for the development of organic farming research. Initial research projects began at the anthroposophic center Goetheanum at Dornach (Ehrenfried Pfeiffer and others) in the 1920s. Organic pioneers like Hans Mueller and Hans Peter Rusch initiated research work on commercial organic farms between 1950 and 1970. In 1973, the private Research Institute of Organic Agriculture Research (FiBL) was founded. From 1973 to 2000, FiBL successfully developed a broad research program with private and state money. Since 2000, the federal state research centers Agroscope started to become involved into organic farming research. Since 2000, the total input of human resources and funding into organic farming research has steadily increased and is currently at approximately 7 million Euros per year. The funding agency for Agroscope and FiBL is the Federal Office for Agriculture (BLW). There are also other sources of funding, like grants from the industry (food retailers, biocontrol companies, alternative veterinary pharmaceutical companies etc.) and from charities.

- In the **UK**, several private pioneer organizations and institutes carried out research work on organic farming: the Soil Association (founded in 1946), the Henry Doubleday Research Association (HDRA, founded in 1954) and the Organic Research Centre, Elm Farm (EFRC, founded in 1981). Since 1991, the government (via DEFRA, The Department for Food, Environment and Rural Affairs) has been funding research; the current funding is almost five million Euros per year; 2000 to 2006 approximately 34 million Euros were spent on organic farming research. Consequently, many universities and state research institutes have become involved in organic farming research. The main objectives of the government funding are transparency for the economic performance of different types of organic farms, evaluation of ecological impacts of organic farming, improvement of the production technique of organic farming and gaining scientific data for amending the standards of organic farming.

Events

In 2007 several events of relevance to the organic sector in Europe and worldwide took place.

- The year started with the successful BioFach Organic Trade Fair, which was accompanied by the BioFach Congress. Numerous papers were presented and discussed. The BioFach Congress 2008 will again present an interesting program and include a whole day on current issues of organic farming in Africa (February 21-24, 2008).

- The 3rd annual scientific congress of the European Research Project Quality Low Input Food QLIF[1] was hosted by the University of Hohenheim (Germany) during March 20 to 23, 2007. The congress was organized in parallel with the 9th Scientific Conference on Organic Agriculture in the German speaking countries, and in total, more than 600 participants joined the organic conferences. The proceedings of the Congress give a good overview of the results of the QLIF project so far, as well as current issues in European organic farming research. In the QLIF project, a range of issues (most importantly to select and breed crop varieties and animal breeds that are better adapted to organic and low input systems) were identified that need to be addressed in future integrated research & development programs (Niggli et al. 2007). The next QLIF Congress will take place in the context of the 2nd ISO-FAR conference in June 2008 in Italy.

- In May 2007, the Food and Agriculture Organization (FAO) organized the International Conference on Organic Agriculture and Food Security in Rome, Italy. At this conference, numerous experts discussed how the food security challenge can be met through organic agriculture. A paper presented at the conference quotes recent models of a global food supply, which indicate that organic agriculture could produce enough food on a global per capita basis for the current world population. "These models suggest that organic agriculture has the potential to secure a global food supply, just as conventional agriculture is today, but with reduced environmental impact," according to the FAO.[2]

- In June 2007, 'Bioacademy' took place in Lednice, Czech Republic, for the seventh time. Apart from an attractive program, meetings of agriculturalists and other technical specialists, most of whom originate from Central and Eastern Europe, took place. The exchange of information and contacts between Eastern and Western Europe has become the main purpose of this conference.

- In August 2007, over 200 participants from 40 countries came together in Schwäbisch Hall, Germany, for the 1st IFOAM International Conference on the Marketing of Organic and Regional Values to discuss ideas, opportunities and strategies to protect organic product identity, traditional knowledge and biodiversity, and thus farmers and rural communities. With over 50 keynote speeches and presentations from experts and leaders in the organic sector, the conference addressed the importance of bringing back value in local and regional economies that are increasingly getting lost in a globalized world.

---

[1] Quality Low Input Food QLIF www.qlif.org
[2] Food and Agriculture Organization FAO, press release of May 3, 2007: Meeting the food security challenge through organic agriculture States should integrate organic agriculture objectives within national priorities, FAO says. Available at www.fao.org/newsroom/en/news/2007/1000550/index.html.
The conference papers are available at www.fao.org/organicag/ofs/index_en.htm

- The year concluded with the European Organic Congress, organized by the IFOAM EU Group, held in December 2007 in Brussels. The aims of the Congress were to assess the current situation of EU organic production and to identify the contribution of organic production to various EU political objectives. According to the IFOAM EU Group, the organic sector demonstrated its strength and maturity at this event, and also proved that organic food and farming has taken center stage in the agriculture policy debate. The presence of a large number of representatives from the European Commission was an indication of the seriousness with which organic farming is now being taken by EU policy makers. The conclusions from the congress suggest that the following steps be taken:

1) to make organic farming a specific focus of the environmental measures to be developed as part of the CAP Health Check;[1]

2) to ensure that some of the proposed increase in rural development funding is targeted at organic farming, in particular in regions and member states that have so far given a low priority to organic farming support;

3) to support the development of a dedicated organic research vision and ideas forum to act as a focus for identifying priorities in research and development and knowledge;

4) improve the involvement of stakeholders in the process of reforming the EU regulation defining organic food and farming;

5) maintain the integrity of organic food with reference to core values to protect against devaluing the concept as the market expands;

6) continue to campaign for GMO policies that that protect the GM free status of organic products and ensure that liability for contamination is consistent with the polluter pays principle; and

7) recognize the urgency for rapid adoption of organic production due to rapidly increasing oil prices and the prospect of oil shortages in the foreseeable future.

In 2008, Europe will host a major organic event, the 16th IFOAM Organic World Congress, to be held in Modena, Italy, from June 16 to 20. The congress will consist of the System Values Track, which highlights the practical experiences of organic farmers, and the second Scientific Conference of the International Society of Organic Agriculture Research ISOFAR; several other parallel and pre-conferences will take place, focusing on organic textiles, viticulture and more. Detailed information is available at the IFOAM and the ISOFAR Homepages (www.ifoam.org and www.isofar.org).

---

[1] According to the European Commission, the so-called 'Health Check' of the CAP aims at streamlining and further modernising the European Union's Common Agricultural Policy. The 'Health Check' of the CAP will build on the approach which began with the 2003 reforms, improve the way the policy operates based on the experience gathered since 2003 and make it fit for the new challenges and opportunities in the EU in 2007. Three main questions will be asked: 1) how to make the direct aid system more effective and simpler; 2) how to make market support instruments, originally conceived for a Community of Six, relevant in the world we live in now; and 3) how to confront new challenges, from climate change, to biofuels, water management and the protection of biodiversity. Detailed information is available at http://ec.europa.eu/agriculture/healthcheck/index_en.htm

References / Further reading

Commission of the European Communities (2004): European Action Plan for Organic Food and Farming (=Commission Staff working document; Annex to the Communication from the Commission {COM(2004)415 final}), Brussels, 2004. Available at http://europa.eu.int/comm/agriculture/qual/organic/plan/workdoc_en.pdf

Lange, Stefan; Williges, Ute; Saxena, Shilpi and Willer, Helga, Eds. (2006) Research in Organic Food and Farming. Reports on organization and conduction of research programs in 11 European countries. Bundesanstalt für Landwirtschaft und Ernährung (BLE) / Federal Agency for Agriculture and Food BLE, Bonn, Germany. Archived at http://orgprints.org/8798/

Llorens Abando, Lourdes and Elisabeth Rohner-Thielen (2007) Different organic farming patterns within EU-25. An overview of the current situation' Statistics in focus 69/2007. http://epp.eurostat.ec.europa.eu/cache/ITY_OFFPUB/KS-SF-07-069/EN/KS-SF-07

Niggli, Urs; Leifert, Carlo; Alföldi, Thomas; Lück, Lorna and Willer, Helga, Eds. (2007) Improving Sustainability in Organic and Low Input Food Production Systems. Proceedings of the 3rd International Congress of the European Integrated Project Quality Low Input Food (QLIF). University of Hohenheim, Germany, March 20 – 23, 2007. Research Institute of Organic Agriculture (FiBL), CH-Frick.http://orgprints.org/10417/

Cropping Patterns in the European Union 2006 (EU 27)

DIANA SCHAACK[1]

More than 6.8 million hectares or 4.1 percent of the usable agricultural area were managed organically in the 27 countries of the European Union in 2006. Forty-one percent or 2.8 million hectares of the organic area is arable land, which is 2.7 percent of all arable land in the EU. Forty-five percent or 3.1 million hectares percent of the organic land is organic permanent grassland – 5.6 percent of all grassland in Europe. Additionally, there are more than 600'000 hectares cultivated with permanent crops – 5.6 percent of all permanent crops in Europe.

Methods

For 15 years, the Central Market and Price Report Office (ZMP) in Germany has observed the organic market in Germany and the surrounding countries in Europe. In a project of the German Federal Organic Farming Scheme, ZMP, together with FiBL and FleXinfo, created a database containing information about land use, animal husbandry, and production and trade volumes. Results are available at the website www.zmp.de/biodaten. The project group is cooperating with the Organic Centre Wales, which has been active in organic data collection for many years (see Lampkin et al. 2007). Most data in this report was gained within the framework of this collaboration, mainly based on Eurostat data and national sources. In some cases, countries had not published detailed land use data at the time of writing of this article, such as for cereals. In these cases, we estimated 2006 data on the basis of information from previous years. The organic shares are generated using Eurostat data, total usable agricultural area from 2006 and land use data from 2005. Eurostat published a report on organic land use and animal husbandry 2005.[2] Land use patterns differ between conventional and organic farming in Europe. In conventional farming in the EU-27, 61 percent of usable agricultural area is arable land, 35 percent is used as permanent grassland and 6 percent are permanent crops. In organic farming, there is far less arable land (41percent), more grassland (43 percent) and a slightly higher share of permanent crops (9 percent). Compared to other continents, grassland and cropland are more evenly spread, but there is nevertheless a lack of arable land for fulfilling the rising demand for organic products in Europe.

The share of arable land, grassland and permanent crops varies widely in the different countries. The countries with the highest share of grassland are the Czech Republic (91 percent), Slovenia (91 percent), Ireland (90 percent) and the United Kingdom (70 percent). Conversely, countries with a high share of arable land include Sweden (80 percent), Estonia (79

---

[1] Diana Schaack, Central Market and Price Report Office ZMP, Organic Agriculture, Rochusstr. 2-6, 53123 Bonn, Germany, www.zmp.de
[2] Different organic farming patterns within EU-25 - an overview of the current situation - Issue number 69/2007; available at http://epp.eurostat.ec.europa.eu/.

percent), Latvia (73 percent) and Lithuania (70 percent). Permanent crops (including grapes) play important roles in the Southern European countries: Cyprus (58 percent), Bulgaria (38 percent), Greece (23 percent), Poland (22 percent), Italy (18 percent) and Spain (17 percent).

Arable land

The most important crops on arable land are **cereals,** with 1.1 million hectares, which is 1.9 percent of the total cereal area in Europe or 16 percent of the organic area. The most important cereal producers are Italy (239'092 hectares), Germany (179'000 hectares), Spain (113'304 hectares, including protein crops) and France (83'861 hectares).

Not all the countries offer detailed information about the different types of cereals; therefore, the organic shares should be viewed as minimum shares. The most important cereal is **wheat, with** at least 400'000 hectares in production. In Italy, 117'686 hectares of durum wheat are grown. Major soft wheat producers are Germany (45'000 hectares), France (30'146 hectares) and the United Kingdom (21'767 hectares). **Rye** is mainly grown in Germany (49'000 hectares), followed by Lithuania (7'402 hectares) and Austria (5'358 hectares). **Barley** is the most important feed grain in conventional production (24 percent of the total cereal area), but it does not play a similarly important role in organic farming, with an estimated 11 percent of the organic cereal production. The main countries producing barley are Italy (32'834 hectares), Germany (20'500 hectares) and Sweden (16'730 hectares). At least 12 percent of organic cereal area (135'000 hectares) is planted with **oats.** Sweden (31'240 hectares), Italy (24'578 hectares), Finland (19'283 hectares) and Germany (18'800 hectares) are the main producers.

Forage (temporary grassland and other forage) is grown on 1.1 million hectares – 16 percent of the organic area and 5.5 percent of total forage production area. Major forage producers in Europe are Italy (297'441 hectares), Germany (122'000 hectares), France (122'513 hectares) and the United Kingdom (102'300 hectares).

Oilseeds play an ancillary roll in organic farming compared to conventional agriculture - less than one percent or 100'000 hectares of European oilseeds are organic. The major countries are France (18'708 hectares), Italy (18'703 hectares) and Romania (16'058 hectares).

Ninety thousand hectares are used for organic **vegetable** production (including strawberries and melons), which is five percent of the total vegetable area in Europe. The main producers are Italy (39'696 hectares), Germany (8'900 hectares), France (8'768 hectares), Spain (5'039 hectares) and the Netherlands (4'584 hectares).

There are about 23'000 hectares of organic **potatoes** in Europe, which is one percent of the total potato area in Europe. The most important producers are Germany (7'500 hectares), Austria (2'426 hectares) and the United Kingdom (2'360 hectares).

Permanent crops

Nine percent of the organic land or 610'000 hectares are used for permanent crops. Half of this area is olive trees (280'000 hectares). Fruit (including berries and nuts) are cultivated on 185'000 hectares. Grapes are grown on 85'000 hectares.

Seven percent of the total **olive** tree area in the 27 EU countries are under organic management. Major producers are Italy (107'233 hectares), Spain (93'432 hectares) and Greece (59'999 hectares).

Six percent of the **fruit** area in Europe is cultivated organically. The main producing countries are Poland (50'200 hectares, including many hectares of new walnut plantations), Spain (49'240 hectares, including 44'600 hectares of nuts), Italy (45'672 hectares) and France (9'179 hectares).

European vineyards have an organic share of 2.6 percent. The main production countries are Italy (37'693 hectares), France (18'808 hectares) and Spain (16'832 hectares).

Grassland

The 3.1 million hectares of organic grassland – 5.6 percent of all grassland in Europe - are mainly in extensive mountainous regions. The meat and milk that are produced there are not always sold through organic sales channels. Particularly in some of the new member states, there is a lack of processing facilities for animal organic products. In some countries, direct payments have encouraged the conversion to organic farming, especially in these extensive areas. The largest grassland areas can be found in Germany (430'000 hectares), the United Kingdom (423'369 hectares) Spain (378'820 hectares), Italy (261'252 hectares) and France (219'763 hectares), although the share of organic grassland in other countries may be much higher (see above).

Further reading

Lampkin, Nicolas, Santiago Olmos, Stephen Lowman and Pauline van Diepen Statistical Report on the Development of Organic Farming in EU-15, Switzerland and Norway 1997-2006 University of Wales, Aberystwyth, November 2007 = Deliverable 5 of the European project Further Development of Organic Farming Policy in Europe with Particular Emphasis on EU Enlargement QLK5-2002-00917

Llorens Abando, Lourdes and Elisabeth Rohner-Thielen (2007): Different organic farming patterns within EU-25. An overview of the current situation. Statistics in focus 69/2007.
http://epp.eurostat.ec.europa.eu/cache/ITY_OFFPUB/KS-SF-07-069/EN/KS-SF-07

The European Market for Organic Food in 2006

SUSANNE PADEL,[1] ALEKSANDRA JASINSKA,[1] MARKUS RIPPIN,[2] DIANA
SCHAACK[3] AND HELGA WILLER[4]

Across Europe, the organic share of the total food market varies from approximately 4.5 percent of total food sales in Switzerland and Denmark, to 3 percent in Germany and approximately 2.5 percent in the UK. In several major European countries, the market for organic food grew substantially between 2005 and 2006, with a growth rate of more than 20 percent in the UK, 18 percent in Germany, 10 percent in Austria and 9 percent in the Netherlands. This trend is expected to have continued in 2007. In countries like Denmark and Switzerland that had experienced stagnation in previous years, the market appears to have entered a renewed stage of growth (Rippin et al. 2007). The rapid growth in demand for organic produce has resulted in supply shortages. In Germany, widespread scarcity of cereals, potatoes, milk, meat and some vegetables varieties led to discernible price increases at both the producer and consumer level. Austria experienced extreme shortages in cereals, potatoes, some fruits and vegetables; even though organic farmland already accounts for 13 percent of the total farmland. Approximately, an additional 10'000 farms would be needed to meet demand. In 2006, Denmark suffered from milk oversupply, but it now is looking for new organic milk producers. In the UK, some companies try to convince farmers to convert to organic by offering financial assistance.

The total value of European retail sales for organic food is estimated to be approximately 14.3 billion Euros.[5] Because of strident efforts to offer the most reliable data possible, this value is similar to the one published in the 2007 edition of 'The World of Organic Agriculture' (Padel & Richter 2007), despite growth reported from several important markets. Products destined for export were eliminated from the calculation in this edition for four countries that are important exporters of organic food and publish such data

The availability of accurate statistics on the market for organic food in Europe remains extremely limited. The values reported here are either published values or the best estimates from a range of experts. The data presented in this section were obtained with different methods.

---

[1] Dr. Susanne Padel and Aleksandra Jasinska, Aberystwyth University, Institute of Rural Sciences, Organic Farming Unit, Llanbardarn Campus, UK, Aberystwyth, SY23 3 AL, UK, www.irs.aber.ac.uk/research/organic_research.shtml
[2] Markus Rippin, Agromilagro Research, Auf der Tränke 17, 53332 Bornheim, Germany, www.agromilagro.de
[3] Diana Schaack, Central Market and Price Report Office ZMP, Organic Agriculture, Rochusstr. 2-6, 53123 Bonn, Germany, www.zmp.de
[4] Dr. Helga Willer, Research Institute of Organic Agriculture (FiBL), Ackerstrasse, 5070 Frick, Switzerland, www.fibl.org
[5] Please note that this figure differs from the figure given by Amarjit Sahota in the chapter on the global organic market in this volume. This is due to a difference in methodology. The data presented by Sahota are based on updates of previous data, based on growth rates communicated by retailers, producers, wholesalers, etc. The data presented here by Padel et al. are based on an expert survey, but the methodology behind the individual figures may differ (see also explanation in the text).

- In a growing number of countries, organic sales of products with bar codes are now monitored through consumer panel services based on representative samples such as AC Nielsen. Such data are reliable to monitor sales through multiple retailers but are less reliable for products that are not barcoded (e.g. fresh fruit and vegetables) and for other outlets, such as farm shops and box schemes.
- Other countries draw on surveys from organic operators in various sectors, or employ estimates based on the value of production from the combination of several approaches.
- Published national data for Austria, Italy and Spain contain the values of products for export, which are shown separately and are not included in the calculation of the total European value. It is not always clear whether the data contain also non-food items that are sold in the same stores.
- Methods of data collection change over time and different sources use different methods even within one country. Comparability between countries and over time remains problematic as a result.
- Estimates for the market value for 2006 were not available for all countries.
- Estimates published in national currencies were converted to Euros using the 2006 average exchange rate from the European Central Bank.[1]

The European Market for Organic Food 2006

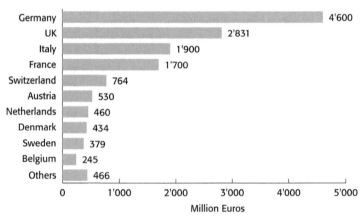

Source: Agromilagro Research, FiBL, IRS/University Wales and ZMP

Figure 20: The European market for organic food

Source: ‚Agromilagro Research, FiBL, Institute of Rural Sciences of Aberystwyth University and ZMP

Germany

Germany remains the largest market for organic products in Europe. In 2006, organic retail

---

[1] Average annual exchange rate of the Euro; see http://sdw.ecb.europa.eu/browse.do?node=2018794

sales increased by 18 percent to 4.6 billion Euros – approximately 2.7 percent of the total food market. This does not include organic food consumption in canteens and catering establishments.

Multiple retailers that were traditionally less important in Germany saw above average growth rates. Specialist organic supermarkets also generated above average growth of 25 percent, while traditional organic shops generally lost market share (BLE 2007a, Hamm & Rippin 2007).

Demand for organic produce has increased faster than supply, and several key products are now undersupplied (Agra Europe). Demand for organic milk in retail shops increased by 35 percent. Starting from comparatively low absolute turnover rates, cakes, biscuits and pastries, with a more than 100 percent increase in sales demonstrated the highest growth rates. Other products with rapidly growing turnovers include cheese curd (over 70 percent), frozen vegetables (over 60 percent), and yoghurts (over 50 percent), followed by butter, pasta, fruit juices, milk and other (Rippin et al. 2007).

A survey of 1000 respondents conducted in the beginning of February 2007 revealed that the number of households that buy organic products regularly increased from 15 percent in 2005 to 22 percent in 2006, and that 55 percent of respondents buy them occasionally (compared to 60 percent in 2005) (BLE 2007b).

UK

The market in the UK grew by an estimated 22 percent in 2006, continuing a strong period of growth over the last few years. In contrast, conversion of farms to organic is slow in most regions (Soil Association 2007).

In 2006, retail sales of organic products in the UK were worth an estimated 1'937 million British Pounds (2.83 billion Euros[1]). This figure includes non-food items, worth approximately 200 million British pounds. Direct retail sales of organic products through mail order and box schemes increased by 53 percent, and consequently the share of direct and independent outlets soared, while the majority of organic foods (75 percent) continue to be sold through multiple retailers.

Sales of organic baby foods in the UK increased by 7 percent, whereas sales of non-organic baby foods declined by 2 percent.

Overall, the organic retail market is worth 2.5 percent of the total food market. In the multiple retailers organic sales accounted for 4.7 percent of all dairy, 4.2 percent of fresh produce and 3 percent of meat.

The organic poultry market continues to increase rapidly, demonstrating 39 percent growth with no signs of abating; the combined sales value of free range and organic eggs exceeded that of cage eggs for the first time. However, the UK's self-sufficiency in organic cereals fell below 50 percent during 2006, further increasing reliance on imported organic grain both for feed and for human consumption. An average of 66 percent of the organic primary

---

[1] Average exchange rate in 2006: 1 Euro =0.68173 British Pounds (GBP)

produce sold by multiple retailers was sourced from the UK, representing no change since 2005. Seventy-nine percent of meat, 96 percent of dairy and eggs and 73 percent of vegetables were sourced from domestic producers, demonstrating higher than average shares, but with demand outpacing supply, imports are likely to increase.

Italy

In 2006, the total Italian organic market reached an estimated value of 2'650 million Euros. Retail outlets sold organic products valuing 1'700 million Euros. Catering outlets account for 200 million Euros and exports for 750 million Euros (Pinton 2007). Sales by specialist shops increased 10 percent, while sales in conventional stores increased 7.5 percent. Sales to processing companies and wholesalers grew by 55 percent and 35 percent respectively, with sales by public procurement, consisting predominately of school meal services, increasing by 19.5 percent. Sales to supermarkets increased by a marginal 2 percent, accounting for approximately 20 percent of country-wide organic sales. Roughly 1'300 new products were introduced, an increase of 30 percent.

Exports of organic products increased between 2005 and 2006 by 25.8 percent, specifically by 146 percent to the UK, 46 percent to Germany, 24 percent to France and 16 percent to Switzerland (AzBio 2007). In multiple retailers, sales of organic oil, sugar, coffee and tea, bread and substitutes, biscuits, sweets and snacks increased in 2006, whereas sales of soft drinks, dairy products, vegetables and fruits (fresh and processed), rice and pasta, baby food and eggs decreased (ISMEA 2006).

France

The retail sales value for 2006 is estimated to be at 1'700 million Euros and continues to grow (Rison Alabert, 2008). Data published in previous years (Padel/Richter 2007) were obtained from a different source, and as a result are not strictly comparable with the Agence Bio data.

The number of consumers purchasing organic products regularly showed a marked increase in France. A survey conducted for Agence Bio (2007) found that 75 percent of consumers regularly shop for organic produce in multiple retailers, 37 percent at weekly markets, 30 percent in specialist shops, 22 percent in delicatessens and 23 percent on farms.

Denmark

The value of the organic market expanded by 18 percent in 2006. Sales through multiple retailers reached 2'701'330'000 Danish Crowns 362 million Euros[1]). This does not include sales through other outlets, which are estimated at approximately 20 percent. In 2006, products worth an estimated 587'094'000 Danish Crowns (78.7 million Euros) were imported, and products worth 275'455'000 Danish Crowns (37 million Euros) were exported (Statistics Denmark 2007). Danish consumers come on third place in terms of per capita spending after Switzerland and Lichtenstein.

Organic retail sales in supermarkets and department stores account for approximately 4.5

---

[1] Average exchange rate in 2006 1 Euro = 7.4591 Danish Krona (DKK)

percent of the total food market (excluding food shops, farm shops and vegetable and fruit markets). The sectors demonstrating the strongest growth were meat and meat products (31 percent), cereals based products (up 24 percent) and baby food (up 23 percent) (Larsen 2007). With a per capita spending of 84 Euros, the Danes are the third highest spenders on organic food after Switzerland and Liechtenstein.

Imports of organic fruit and vegetables increased from 16 million Euros in 2004 to 22 million Euros in 2005, while exports decreased from 5 million in 2004 to 4 million Euros in 2005. Exports of organic meat, especially pork, increased by 10–15 percent in 2006, following a 50 percent increase in 2005. The lower growth rate was due to supply shortages, which are expected to continue during 2007, despite a predicted increase in production of 50 percent (Agra Europe Weekly).

Austria

In 2006, domestic organic retail sales were worth an estimated 530 million Euros - an increase of more than 10 percent compared to 2005. Multiple retailers account for 65 percent of sales, followed by 14 percent through specialty outlets, 6 percent through canteens and catering and 5 percent through direct sales. Products worth 60 million Euros are exported (Bio-Austria, 2007).

Organic retail sales now account for 5.4 percent of the total food market (Klingbacher 2007). For certain product categories, shares are considerably higher, for example - 21 percent of all egg sales in multiple retailers, 14 percent of all milk (fresh and extended shelf life) and 9 percent of all yogurt sold is organic (RollAMA /AMA).

Since 2003, 50 percent of the food and drink in nurseries, 30 percent in hospitals and 17 percent in nursing homes has been organic (ZMP Ökomarkt Forum 2007).

The Netherlands

Dutch consumer spent about 460 million Euros on organic food in 2006, an increase of 9 percent compared to 2005. Organic retail sales now account for 1.9 percent of the total food and drink market. In the fresh product sector, organic products have 2.8 percent of the market (Biologica 2007).

The most important sales channels are whole food shops with a 43 percent share of the total organic turnover, but discounters gained in importance with a 26 percent increase in organic sales (Biologica 2007).

Fresh organic dairy products (excluding cheese and butter) have the greatest share of the total food market (3.8 percent), with a total sales value of 77 million Euros, followed by fresh fruit and vegetables (2.8 percent). Although sales for organic fruit and vegetables (and potatoes) increased from 140.2 million Euros to 149.5 million Euros, the market share dropped due to higher prices, with consequently a higher turnover for conventional fruit and vegetables. Sales of organic meat totaled 57.6 million Euros, while bread was worth 29 million Euros (Biologica 2007).

Spain

The total turnover of organic products was estimated to be worth 350 million Euros in 2006, the vast majority of produce destined for export markets. In previous years, it was not possible to distinguish between products for export and for the domestic market, thus a figure for the total amount was published. There still are no published statistics for the national market, but the value has been estimated at approximately 70 to 80 million Euros (Gonzálvez 2007).

The Ministry of Agriculture has set aside 36 million Euros to develop the organic food sector in Spain over the next four years (Agra Europe Weekly).

Switzerland

The market value was estimated to be worth 1.202 billion Swiss Francs (764 million Euros) in 2006[1]. Swiss consumers spent an average of 102 Euros on organic products in 2006 – more per capita than in any other country in the world. Organic products now account for 4.5 percent of the total food market. After consolidation in 2005, the organic turnover increased again in 2006 by 1.6 percent (Bio Suisse 2007) and growth is expected to have continued in 2007.

Belgium

Turnover of organic food sales rose to 245 million Euros in 2006 (Puur 2007) which is an increase of 33 percent compared to 2005. More than 60 percent of the turnover was generated via traditional supermarkets, and 31 percent of organic food was sold via specialized organic shops. Market share of eggs was 7 percent, rice 4 percent, vegetables 3 percent, bread 2.3 percent, fruit 1.6 percent and of potatoes 1 percent.

Sweden

The revenue from sales of organic products has been estimated to be 3.51 billion Swedish Crowns (379 million Euros[2]) in 2006 or between 2 to 3 percent of the total sales of food in Sweden (van der Krogt, 2007).

Finland

In 2006, the retail sales of organic products were around 57 million Euros, an approximately 0.8 percent share of the total food market (Heinonen 2007). One major supermarket chain withdrew from selling organic products and as a result, the turnover of organic food decreased substantially. There has been a fall in sales of most vegetables, fresh bread and other grain products. However, at the same time there has been a remarkable increase in sales of, for example, eggs and vegetable oils.

---

[1]Average exchange rate in 2006 1 Euro = 1.548 Swiss Francs (CHF)
[2] Average exchange rate in 2006 1 Euro = 9.2544 Swedish Krona (SEK)

Norway

The retail sales value for organic food is estimated to be worth 518'650'465 Norwegian Crowns (64.5 million Euros[1]). The share of organic sales of the total food market rose considerably during 2006, and now accounts for 0.78 percent of the total food market. The shares are highest for baby food (nearly 10 percent), followed by eggs (1.96 percent), milk and dairy products (1.67 percent) and bread and cereal based products (1.32 percent) (Norwegian Agricultural Authority 2007)

A report by the Centre for Rural Research (Agra Europe Weekly, December 2006) found that 1.5 percent of Norwegian producers plan to convert to organic production, compared to 4 percent in 2002. Norway's dairy giant, Tine, is failing to keep up with demand for organic milk, while a major ice cream manufacturer, Hennig-Olsen, planned to launch a range of organic ice cream, but was forced to look abroad for supplies of organic milk.

The Czech Republic

In 2006, organic retail sales were worth 760 Czech Crowns (26.8 million Euros[2]), an increase of 49 percent since 2005. Organic retail sales currently account for just 0.35 percent of the total market. However, this is predicted to increase to one percent by 2010 (Agra Europe Weekly 2007, Fresh Plaza Place 2007).

Poland

More and more supermarket chains are entering the organic market or are expanding their organic product range, including the Polish company Organic Farma Zdrowia and international corporations such as Carrefour and Tesco. Increased consumer demand for organic products has resulted in a lack of organic raw ingredients for processing (NzW 2007).

References

Agence Bio (2007) Marché alimentaire bio en France: une croissance de près de 10 percent par an. Press release. Agence Bio, http://www.agencebio.fr/pageEdito.asp?IDPAGE=59&n2=67

Agra Europe Weekly (2007): various issues.

AZBIO (2007), n. 1 / 2

Bio-Austria (2007) Umsatz Bio-Lebensmittel in Österreich 2006. http://www.bio-austria.at

Bio Suisse (2007): Bio im Neuaufbruch. Bio Suisse press release of April 3, 2007. Basel, 2007, http://www.biosuisse.ch/media/de/pdf2006/d_texte_bio_suisse_07.pdf

Biologica (2007): Bio-Monitor 2006, Biologica, Utrecht. http://www.biologica.nl/docs/200704031708078385.pdf Jaarreport NL

BLE - Bundesanstalt für Landwirtschaft und Ernährung BLE (2007b): Ökobarometer 2007: Aus Gelegenheitskäufern werden regelmäßige Bio-Kunden Supermärkte machen Neukunden auf Bio aufmerksam. Press release of Bundesanstalt für Landwirtschaft und Ernährung BLE / Bundesprogramm Ökologischer Landbau, Bonn, Berlin. http://www.oekolandbau.de/fileadmin/redaktion/dokumente/journalisten/Pressemeldungen/Text001_Oekobrometer2007.pdf.

BLE - Bundesanstalt für Landwirtschaft und ErnährungBonn/Berlin (2007a): Pressemitteilung Bio-Markt 2006:

[1] Average exchange rate in 2006 1 Euro= 8.0472 Norwegian Krona (NOK)
[2] Average exchange rate in 2006 1 Euro =28.34 Czech Crowns (CZK)

Hohes Marktpotenzial steht gegen stagnierendes Wachstum im Ökolandbau, Verbraucherinteresse an Bio-Produkten ist ungebrochen groß. Pressemitteilung Bundesanstalt für Landwirtschaft und Ernährung BLE / Bundesprogramm Ökologischer Landbau, Bonn, Berlin. http://www.bundesprogramm-oekolandbau.de/fileadmin/sites/default/files/Presse/boel_bio-marktzahlen2006.pdf

EkoConnect (2007). Infoletter Nr. 13 newsletter (2007): No 13, http://www.ekoconnect.org/pdf/Infobrief_13/Infobrief-englisch-13.pdf. www.ekoconnect.org

Fresh Plaza (2007) Czech Organic Market will quadruple by 2011. May 24, 2007, www.freshplaza.com/news_detail.asp?id=1850

Gonzálvez, Victor / Spanish Society of Organic Agriculture SEAE, written communication of December 17, 2007.

Hamm, U., Rippin, M. (2007): Marktdaten aktuell: Öko-Lebensmittelumsatz in Deutschland 2006. Öko-Lebensmittelumsatz in Deutschland. http://www.agromilagro.de/resources/UmsatzOeko2006.pdf

Heinonen, Sampsa / Evira (2007): Written Communication

Institute of Rural Sciences (IRS), University of Wales: Eurodata 2007. IRS University of Wales, Aberystwyth; unpublished

ISMEA Osservatorio biologico ISMEA newsletter (2006): Newsletter No 3. www.ismea.it

Klingbacher, Elisabeth/Bio Austria, Vienna, written coomunication of September 11, 2007

Larsen, Paul Henning / Statistics Denmark (2007). Personal communication.

Norwegian Agricultural Authority (2007): Produksjon og omsetning av økologiske landbruksvarer Rapport for 2006. Norwegian Agricultural Authority, Report No. 10/2007. http://www.slf.dep.no/iKnowBase/Content/6316/PRODUKSJON%20OG%20OMSETNING%20AV%20%c3%98KOLOGISKE%20LANDBRUKSVARER%202006.PDF

NzW 'Notizen zur Wettbewerbslage' 4/2007. Monatsschrift des Absatzförderungsfonds der deutschen Land- und Ernährungswirtschaft (CMA, ZMP)

Padel, Susanne and Toralf Richter (2007): The European Market for Organic Food 2005. In: Willer/Yussefi (2007). The World of Organic Agriculture 2007. IFOAM Bonn and FiBL Frick

Pinton, Roberto / Pinton Organic Consulting: Written Communication of November 20, 2007

puur, Bio in Cijfers November 2007. Bioforum. Berchem

Rippin M., Kasbohm A., Engelhardt H., Schaack D., Hamm U. (2007): Ökomarkt Jahrbuch 2007. Verkaufspreise im ökologischen Landbau 2005/2006. ZMP, Bonn. http://www.zmp.de/shop/mzm/oekomarkt_jahrbuch_mzm_68.asp

Rison Alabert, N (2008) Personal communication.

Soil Association (2007): Organic Market Report. The Soil Association. Bristol. Info at http://www.soilassociation.org/marketreport

Statistics Denmark (2007), Table OEKO4: External trade with organic products by imports and exports and commodities; http://www.statbank.dk/statbank5a/SelectVarVal/Define.asp?Maintable=OEKO4&PLanguage=1

RollAMA/AMA Marketing: Various organic market statistics at http://www.ama-marketing.at/home/groups/7/Marktentwicklung_Bio.pdf

van der Krogt (2008) Växanda Marknad 2007 (forthcoming) Ekologiska Lantbrukarna, Uppsala www.ekolantbruk.se

ZMP Ökomarkt Forum (2007): Various issues. ZMP, Bonn, www.zmp.de

Table 21: The European market for organic food 2006

Country	Per capita consumption (€/person)	Turnover (Mio €) Exports excluded if separate figure available	Exports (Mio €)	Share of total domestic market (percent)
Austria	64	530	60	5.4
Belgium	23	245	No data	1.7
Bosnia Herzegovina	No data	No data	1	No data
Croatia	5	20	No data	0.4
Czech Rep.	3	26.8	No data	0.35
Denmark	80	434	37	5.0
Finland	11	57	No data	0.8
France	27	1'700	No data	1.1
Germany	56	4'600	No data	2.7
Greece (2005)	5	50	No data	No data
Hungary	1	7	No data	No data
Ireland (2005)	16	66	No data	No data
Italy	32	1'900	750	No data
Liechtenstein	86	3	No data	No data
Netherlands	28	460	No data	1.9
Norway	14	64.5	No data	0.8
Poland	1	20	No data	No data
Portugal (2005)	5	50	No data	No data
Romania	0.1	3	No data	No data
Slovak Republic[1]	1	4.3	No data	No data
Spain	2	70	280	0.1
Sweden	42	379	No data	2-3
Switzerland	102	764	No data	4.5
UK	47	2'831	No data	2.5
Ukraine	No data	4	No data	No data
Total*	29	14'309	No data	No data

* Includes estimate for missing countries

Source: Survey by Aberystwyth University/Institute of Rural Sciences, Agromilagro Research, FiBL and ZMP

Per capita consumption calculated by Susanne Padel, Aberystwyth University/Institute of Rural Sciences

For information on data providers / sources please see the end of the European chapter.

[1] Original Currency: 160'000'000 Slovakia Koruny (SKK); Average exchange rate in 2006 1 Euro = 37.234 SKK

Trends in the Organic Retailing Sector in Europe 2007[1]

TORALF RICHTER[2]

The specialist organic retailing sector[3] in Europe profited from the positive organic food market developments in 2007. Retailers specializing in organic products set trends for the development of the entire organic market, boosting sales and injecting crucial momentum into the market. Presently, the organic retail sector in Europe represents approximately 5 to 25 percent of the total organic market; the importance of the specialist organic retail sector varies from country to country.

In most countries, the organic retail sector grew proportionately to the general organic market. In some countries, such as in Italy, Switzerland and the Netherlands, retail growth outpaced growth in production. Major European countries with large organic retailing sectors include Germany, France, Italy, the Netherlands, Austria and Switzerland. In addition, the organic retail sector has gained more significance in some Southern and Eastern European and Scandinavian countries. New enterprises or branches of specialist organic retailers emerge on a nearly constant basis in Europe.

The diversification of sales concepts and marketing tools are visible indications that the organic retailing sector is maturing. In the past, the organic retail sector was characterized by smaller shops located in the periphery of city centers. In recent years, large organic supermarkets, organic convenience stores in train stations or shopping centers, organic discounters, natural cosmetic shops and shops dedicated to specific goods, such as wine or tea, have contributed to the rapid diversification of retail outlets.

A further trend is the rising participation of organic retailers in marketing programs, such as [ECHT BIO.] (see picture) in Germany and Switzerland or 'b'io' in Italy. The marketing programs, which are driven by organic wholesalers, include promotional materials for the point of sale or common price promotion activities. Approximately 20 percent of all German and Swiss organic shops have joined the marketing program [ECHT BIO.]. Approximately

---

[1] The information presented is mainly based on market intelligence sources like Bio-Markt.Info (www.organic-market.info), Biopress (www.biopress.de), Organic Monitor (wwworganicmontor.com) and the BioFach-Newsletter (www.biofach.de/en/newsletter/archive)as well as on personal contacts within the organic retailer sector.
[2] Dr. Toralf Richter, Bioplus AG, Seon, Switzerland, http://www.biopartner.ch/bpl/News00_461.asp. Toralf Richter is Vice President of the Organic Retailer Association ORA. Information is available at www.o-r-a.org/
[3] There is no international common standard for the organic retail sector. The Organic Retailers Association (ORA) defines organic retailers as specialized organic food shops, natural and health food shops, organic supermarkets, natural drugstores, organic farm shops and all independent retailers with an explicit organic product focus. Organic retailers prioritise organic products in their assortment decisions and communication policy. Their sales staff has thorough knowledge of organic products and organic farming.

200 specialist retailers participate in the program 'b'io' in Italy.

Fresh and frozen food and natural care and cosmetic products are the focus of assortment trends in the organic retail sector.

The international shortage of raw materials represents a challenge for the coming years, and will likely result in price inflation. In order to assure continued prosperity, organic retailers must strive to meet consumer needs and stem the move of consumers to conventional retailers marketing organic products; this process must be linked to a search for more central and profitable store locations.

Germany

The organic retail sector in Germany has been shaped by the increasing interaction with conventional marketing structures. The national retailer REWE, with 3'000 conventional supermarkets throughout Germany, for example, has begun to explore the organic retail sector, and has opened two successful organic supermarkets, with plans for further expansion. Furthermore, various traditional and exclusive manufacturers of the organic retailing sector left the organic retailing supply chain, choosing to sell their product assortments directly to selected conventional retailers. Even the association Demeter, whose biodynamic products are sold worldwide almost exclusively via specialist organic retailers, increased its distribution channels by delivering Demeter products to selected conventional retailers. It is estimated that approximately ten percent of all 'Demeter' products in Germany are sold via the conventional retail trade. In 2007, the conventional discount group Lidl & Schwarz announced its intention for the acquisition of a significant share in the organic supermarket chain Basic. However, due to negative publicity and consumer and supplier boycotts, the planned sale of stock from Basic was halted.

Switzerland

In Switzerland, organic retails chains have been largely absent due to the market domination by cooperative retailer Coop and Migros. Recently two chains of specialized health food shops emerged - Mueller and Egli. Other companies, such as the organic supermarket Yardo and a chain of natural cosmetic shops, are in early stages of development. In 2007, three leading wholesalers - Eichberg, Vanadis and Via Verde - merged to create the new company Bio Partner Schweiz AG. The new company holds a market share of approximately 80 percent of organic retailers in Switzerland.

Austria

The entry of German organic supermarket chains Basic and Dennree into the Austrian retail sector generate growth in the sector, while the Austrian enterprise Livit continued expansion. Marking an exciting development for the organic retail sector, the Austrian manufacturer Sonnentor began selling franchises for organic tea, spice and herb concept stores in Austria, Germany and Switzerland, after a successful trial period of pilot shops in Austria.

Italy

While the organic food turnover in the conventional retail sector is stagnating, the organic retail sector experienced a 15 percent increase in sales over the last year. The largest organic wholesaler Ecor operates its own organic supermarket chain called Natura Si, with 60 outlets nationwide. Ecor also owns the marketing program b'io, which reaches 206 shops. Ecor exemplifies the general European tendency for smaller regional wholesalers to merge into consolidated national wholesalers. The organic wholesalers in Italy profited from the permanent increase of meals served in school canteens (approximately 650 schools with organic meals).

The Netherlands

After a phase of steady organic market growth of between five and seven percent, there is evidence that increasingly rapid growth can be expected in the coming years. New business initiatives from wholesalers and individual health food store owners drive market growth. In 2005, EkoPlaza launched an entirely new business concept of large organic supermarkets with a surface area of 1'500 m^2. This market launch initiated a wave of other market launches around the country, and spurred efforts rethink existing shop concepts.

United Kingdom

Whole Foods Market, the leading US retailer of organic food, entered the European market in 2007. In central London, Whole Foods opened a huge warehouse style supermarket with a surface of 7'400 m^2 and 500 employees; approximately one third of the product assortment is certified organic.

Europe: Organic Farming Statistics

Table 22: Organic agricultural land and farms in Europe 2006

Country	Organic agricultural area (ha)	Share of total agricultural area	Farms
Albania	1'000	0.1%	100
Austria	361'487	13.0%	20'162
Belgium	29'308	2.1%	783
Bosnia Herzegovina	726	0.0%	329
Bulgaria	4'692	0.2%	218
Croatia	6'204	0.2%	368
Cyprus	1'979	1.3%	305
Czech Rep.	281'535	6.6%	963
Denmark	138'079	5.3%	2'794
Estonia	72'886	8.8%	1'173
Finland	144'558	6.4%	3'966
France	552'824	2.0%	11'640
Germany	825'539	4.8%	17'557
Greece	302'256	7.6%	23'900
Hungary	122'765	2.9%	1'553 (2005)
Iceland	5'512	0.4%	27
Ireland	39'947	0.9%	1'104
Italy	1'148'162	9.0%	45'115
Latvia	118'612 (2005)	7.0% (2005)	4'095
Liechtenstein	1'027	29.1%	41
Lithuania	96'718	3.5%	1'811 (2005)
Luxemburg	3'630	2.8%	72
Macedonia	509	0.0%	101
Malta	20	0.2%	10
Moldova	11'405	0.5%	121
Montenegro	25'051	4.8%	15
Netherlands	48'424	2.5%	1'448
Norway	44'624	4.3%	2'583
Poland	228'009	1.5%	9'187
Portugal	269'374	7.3%	1'696
Romania	107'582	0.8%	3'033
Russian Federation, European Part	3'192	0.0%	8
Serbia	906	0.0%	35
Slovak Republic (Mid 2007)	121'461	5.8%	279
Slovenia	26'831	5.5%	1'953
Spain	926'390	3.7%	17'214
Sweden	225'385	7.1%	2'380
Switzerland	125'596	11.8%	6'563
Turkey	100'275	0.4%	14'256
UK	604'571	3.8%	4'485
Ukraine	260'034	0.6%	80
Total	7'389'085	1.6%	203'523
Total EU 27	6'803'024	4%	178'896

Source: Survey by FiBL in cooperation with ZMP and the Institute of Rural Sciences of Aberystwyth University, 2008

Table 23: **Use of organic cropland in Europe 2006**

Please note that this table refers to the whole of Europe whereas the article by Schaack is about cropping patterns in the European Union.

Main land use type	Crop category	Agricultural area (ha)
Arable land	Arable crops, no details	139'208
	Cereals	1'295'568
	Fallow land as part of crop rotation	148'584
	Flowers and ornamental plants	270
	Green fodder from arable land	1'068'583
	Industrial crops	11'543
	Medicinal & aromatic plants	28'958
	Oilseeds	97'597
	Other arable crops	250
	Protein crops	115'044
	Root crops	29'780
	Seeds and seedlings	14'481
	Textile fibers	18'912
	Vegetables	93'062
Arable land		*3'061'840*
Permanent crops	Citrus fruit	25'001
	Fruit and nuts	266'842
	Grapes	95'130
	Industrial crops	83
	Medicinal & aromatic plants	96
	Nurseries	550
	Olives	289'006
	Other permanent crops	14
	Permanent crops, no details	23'607
	Tropical fruit and nuts	773
Cropland, other/no details		*82'381*
Permanent grassland		*3'171'533*
Other		*269'850*
No information		*102'377*
Total		7'389'085

Source: Survey by FiBL in cooperation with ZMP and the Institute of Rural Sciences of Aberystwyth University, 2008

Table 24: **Certified wild collection in Europe 2006**

Country	Use of wild collection area	Organic wild collection area (ha)
Albania	Wild collection, no details	12'700
Bosnia Herzegovina	Wild collection, no details	440'000
Bulgaria	Medicinal & aromatic plants	11'014
Croatia	Bee pastures	17'662
	Medicinal & aromatic plants	10'000
Finland	Fruit and berries	7'507'523
Iceland	Aquatic products	200'000

Country	Use of wild collection area	Organic wild collection area (ha)
	Wild collection, no details	8'566
Macedonia	Wild collection, no details	1'592
Montenegro	Wild collection, no details	133'800
Romania	Wild collection, no details	38'700
Serbia	Fruit and berries	165'000
	Mushrooms	275'000
	Vegetables	500'000
	Wild collection, no details	162'000
Turkey	Wild collection, no details	92'514
Ukraine	Fruit and berries	4'200
	Mushrooms	800
	Wild collection, no details	13'000
Total		9'594'071

Source: FiBL Survey 2008

Information sources / Data providers for land area, land use and for market data

- Albania: Data provided by Anula Guda, Sasa, Tirana, Albania
- Austria. Land area: Source: Eurostat: Organic crop area 2006, ec.europa.eu/eurostat; Farms: Data provided by Elisabeth Klingbacher, Bio Austria, Source: Grüner Bericht 2007.
  Market data provided by Elisabeth Klingbacher, Bio Austria, Vienna, Source: AC Nielsen.
- Belgium: Source: Eurostat: Organic crop area 2006 / Number of registered organic operators 2006. ec.europa.eu/eurostat
  Market data provided by Petra Tas, Bioforum, Source: puur, Bio in Cijfers November 2007. Bioforum. Berchem
- Bosnia Herzegovina: Data, including export figure, provided by / Source: Maida Hadziomerovic, Organska Kontrola (OK), Sarajevo. Bosnia & Herzegovina
- Bulgaria: Total organic land according Eurostat, Organic crop area 2006 ; Land use details & number of farms provided by Stoilko Apostolov, Bioselena, Karlovo, Bulgaria, Statistic data of organic farming in Bulgaria up to 31st of December, 2006
- Croatia: Data, including market data, provided by Darko Znaor, Independent Consultant, Tuskanac 56B, 10000 Zagreb, Croatia and Sonja Karloglan Todorović, Ecologica, Vlaska 64, 10000 Zagreb, Croatia
- Cyprus: Land area: Eurostat: Organic crop area 2006 / Number of Farms as of 2005; provided by Ionanis Papastylianou, Agricultural Research Institute, Nicoisa, Cyprus.
- Czech Rep.: Data provided by Tom Vaclavik, Green Marketing, Source: Ministry of Agriculture, CZ-Prague: Market data provided by Tom Vaczlavik, Green Marketing, Moravské Knínice, Czech Republic
- Denmark: Source: Eurostat: Organic crop area 2006 / Number of registered organic operators 2006. ec.europa.eu/eurostat.
  Market data: Denmark: Statistics Denmark (2007), Data provided by Poul Henning Larsen, Statistics Denmark, see also various tables at http://www.statbank.dk
- Estonia: Source: Eurostat: Organic crop area 2006 / Number of registered organic operators 2006. ec.europa.eu/eurostat

- Finland: Evira 2007, Organic farming 2006 - Statistics, Loimaa
  Market data provided by Heinonen, Sampsa / Evira, Source: ACNielsen Finland, Consumer Panel 2006
- France: Source, Agence Bio, Les chiffres de l'agriculture biologique en 2006, www.agencebio.fr
  Market data provided by Nathalie Rison Alabert, Agence Bio
- Germany; Source: ZMP Bio-Strukturdaten 2006, ZMP, Bonn, 2007, www.zmp.de
  Market data: Source: Hamm, U., Rippin, M. (2007): Marktdaten aktuell: Öko-Lebensmittelumsatz in Deutschland 2006. Öko-Lebensmittelumsatz in Deutschland. Available at http://www.agromilagro.de/resources/UmsatzOeko2006.pdf
- Greece: Data provided by Nicoletta van der Smissen, Source: Ministry of Agriculture, Farms according to Eurostat.
  Market data provided by Nicoletta van der Smissen, DIO, Greece
- Hungary: Eurostat: Organic crop area 2006. Number of farms as of 2005. Land use data according to Biokontroll Hungary: Report of the Activities 2006, http://www.biokontroll.hu/eves/2006/2006beszamolo_eng.pdf
  Market data: Hungary: Estimate Susanne Padel, Aberystwyth University, Institute of Rural Sciences
- Iceland: Data provided by / Source: Gunnar Gunnarsson, Vottunarstofan Tún ehf., Reykjavík, Iceland
- Ireland: Data according to Teagasc at http://www.teagasc.net/publications/2007/20070131/ntc2007paper07.htm; Land use data: Estimates by Diana Schaack, ZMP, Bonn, Germany
  Market data provided by Eddie MC Auliffe, Organic Unit, Department of Agriculture
- Italy: Data provided by SiNAB: L'agricoltura biologica in Cifre al 31/12/2006. http://www.sinab.it/allegati_news/458/bio_in_cifre_2006_DEF_con_grafici.pdf. Source: MIPAF; Market data provided by Roberto Pinton, Pinton Organic Consulting, Padova, Italy
- Latvia: Source: Eurostat: Organic crop area 2005 / Number of registered organic operators 2006. ec.europa.eu/eurostat
- Liechtenstein: Data (including market data) provided by Klaus Büchel Anstalt, Mauren, Liechtenstein
- Lithuania: Source: Eurostat: Organic crop area 2006 / Number of registered organic operators 2005; ec.europa.eu/Eurostat
- Luxemburg: Data provided by / Source: Administration des services techniques de l'agriculture (ASTA), Monique Faber, Luxemburg
- Macedonia: Data provided by / Source: Radomir Trajković, PROBIO and Balkan Biocert Skopje, Macedonia
- Malta: Eurostat: Organic crop area 2006 / Number of registered organic operators 2006
- Moldova: Data provided by Jens Treffner and Lutz Mammel, Ecconnect, Dresden, Germany. Farms as of 2005. provided by Iurie Senic EcoProdus - Organic Farmers Association, Chisinau, Moldova
- Montenegro: Data provided by Lina Albitar, IAMB Bari, Survey of the Mediterranean Organic Agriculture Network (MOAN); Bari, Itay
- Netherlands: Source (including market data): Biologica (2007): Bio-Monitor 2006, Biologica, Utrecht. http://www.biologica.nl/docs/200704031708078385.pdf

- Norway: Eurostat: Organic crop area 2006 / Number of registered organic operators 2006; ec.europa.eu/Eurostat
  Market data provided by Matthias Koesling, Bioforsk, Tingvoll, Norway, Source: Statens Lant-bruksvorwaltning/Norwegian Agricultural Authority (2007): Produksjon og omsetning av økologiske landbruksvarer Rapport for 2006. Norwegian Agricultural Authority, Report No. 10/2007. http://www.slf.dep.no/iKnowBase/Content/6316/PRO-DUKSJON%20OG%20OMSETNING%20AV%20%c3%98KOLOGISKE%20LANDBRUKSVARER%202006.PDF
- Poland: Data provided by Dorota Metera, Bioekspert, Warzawa, Poland, written communication,, Source: GIJHARS
  Market data: Estimate provided by Tom Vaclavik, Green Marketing/Organic Retailer Association, Moravské Knínice, Czech Republic
- Portugal: Data provided by Ana Firmino, University of Lisbon. Source: Ministry of Agriculture, Fileira Agro-Alimentar
  Market data according to BioFach, provided by Toralf Richter, FiBL (now Biopartner, Seon, Switzerland), November 27, 2006
- Romania: Land area / land use: Eurostat: Organic crop area 2006; Producers: Source: Ministry of Agriculture and Rural Development MADR, Producers List 2006, Download at http://www.maap.ro/pages/page.php?self=01&sub=0107&tz=010707&lang=2
  Market data provided by Stefan Simon, Ekoconnect, Dresden, Germany
- Russian Federation: Source/Data provided by Andrey Khodus, Eco Control
- Serbia: Data provided by Lina Al Bitar, IAMB Bari, Survey of the Mediterranean Organic Agriculture Network (MOAN); Bari, Itay
- Slovak Republic: Data provided by Zuzana Lehocka, Research Institute of Plant Production, Pieštany, Slovak Republic, Source: UKSUP
  Market data estimate provided by Tom Vaclavik, Green Marketing/Organic Retailer Association, Moravské Knínice, Czech Republic
- Slovenia: Eurostat: Organic crop area 2006 / Number of registered organic operators 2006
- Spain: Source: MAPA, Madrid, Spain, http://www.mapa.es/alimentacion/pags/ecologica/pdf/2006.pdf
  Market data: Estimate by Victor Gonzálvez, Spanish Society of Organic Agriculture SEAE, Catarroja, Spain
- Sweden: Source: Statistics Sweden, Statistical Yearbook of Sweden 2007, Chapter 11, http://www.scb.se/statistik/_publikationer/OV0904_2007A01_BR_A01SA0701.pdf, translation of land use details by ZMP, Bonn, Germany
  Market data: Data provided by Ulrika Geber, Swedish University of Agricultural Sciences SLU, Centre for sustainable agriculture CUL, Uppsala, Sweden; Source: van der Krogt, Dirk.2007. Växande Marknad. Ekologiska lantbrukarna. Uppsala, Sverige. www.ekolantbruk.se
- Switzerland: Total farms and area provided by Christine Rudmann, Research Institute of Organic Agriculture (FiBL), Frick Switzerland, based on data from the certifiers. Land use details according to Bundesamt für Statistik, BFS, Neuchatel, Switzerland
  Market data: Bio Suisse (2007): Bio im Neuaufbruch. Bio Suisse press release of April 3, 2007. Basel, 2007, http://www.biosuisse.ch/media/de/pdf2006/d_texte_bio_suisse_07.pdf
- Turkey: Data provided by /Source: Erdal Süngü, Ministry of Agriculture MARA, Ankara
- UK: Source: Eurostat: Organic crop area 2006, Operators 2006. ec.europa.eu/eurostat
  Market data: UK: Soil Association (2007): Organic Market Report. The Soil Association. Bristol. Info at http://www.soilassociation.org/marketreport
- Ukraine: Data (including market data) provided by Eugene Milovanov, Organic Federation of Ukraine, Kiev, Ukraine

Organic Farming in the Mediterranean Region: Towards Further Development

LINA AL-BITAR[1]

Introduction

The Mediterranean Agronomic Institute of Bari (IAMB),[2] an international intergovernmental organization promoting the development of agriculture in the Mediterranean region, initiated data collection on the organic sector in the Mediterranean area in 1999.

Since 2005, a separate chapter in 'The World of Organic Agriculture' has been dedicated to the Mediterranean area, and IAMB has been committed to supporting the global organic survey in this region.

The Mediterranean region includes the Mediterranean countries of the European Union (Cyprus, France, Greece, Italy, Malta, Portugal, Slovenia and Spain), the Eastern Mediterranean countries (Albania, Bosnia and Herzegovina[3], Croatia, Macedonia, Montenegro and Serbia), the Middle East or Mashrek (Egypt, Jordan, Israel, Lebanon, Palestine, Syria and Turkey) and North Africa or Maghreb (Algeria, Libya, Morocco and Tunisia).

Historical development of organic agriculture

Organic agriculture is not a novelty in the Mediterranean. On the contrary, in many countries, organic agriculture was introduced 30 to 40 years ago – at first in the EU Mediterranean countries by pioneer organizations and associations, and slightly later in some countries of the Eastern and Southern shores (Israel, Egypt, Morocco, Tunisia and Turkey). The main drivers were foreign private companies looking for new investment opportunities stimulated by the growing market for organic food in Europe. However, the insufficiency or the entire absence of policies supporting the organic sector have slowed down its development in the Mediterranean area. Earnest development of the organic movement has begun over the course of the last decade, and today organic farming is present in most Mediterranean countries (Al Bitar, 2004). The table illustrates some important steps in the evolution of the organic sector in the Mediterranean Region.

---

[1] Dr. Lina Al Bitar, Mediterranean Agronomic Institute of Bari IAMB, Via Ceglie, 9, 70010 Valenzano (BA), Italy, www.iamb.it

[2] IAMB is the Italian seat of CIHEAM, the Centre International de Hautes Etudes Agronomiques Méditerranéennes/International Centre for Advanced Mediterranean Agronomic Studies.

[3] Including the Federation of Bosnia and Herzegovina and Srpska Republic

Milestones in the history of organic agriculture in the Mediterranean

1960s-1970s	Organizations and associations promoting organic farming in EU Mediterranean countries proliferate.
1977	Organic farming in Egypt begins; the Sekem foundation leads the way, applying German experiences.
1980s	After ten years of unsuccessful attempts, organic farmers in Israel found the Israeli Bio Organic Association (IBOAA), setting the stage for the development of the sector in Israel.
1984-1985	A consequence of the growing market for organic products in Europe, development of organic sector in Turkey flourishes.
1980s	Organic agriculture begins to thrive in Tunisia, through private initiatives.
1986	Beginning of the organic movement in Morocco, initiated by citrus growers with French support.
End 1980s 1990	Croatian and Serbian endeavors in organic agriculture are instigated.
1990	The Mediterranean group on organic agriculture - AgriBioMediterraneo - is established; it becomes the IFOAM Mediterranean group in 1997. The Egyptian Biodynamic Association (EBDA) is founded, giving birth to the Centre of Organic Agriculture in Egypt (COAE), a local certification body.
1990s	The move towards organic agriculture is set in motion in Lebanon as a reaction on the excessive use of chemicals.
1992	ETO, the Turkish association for organic agriculture, is born.
1994	Turkey establishes its own regulation.
1997	The Organic Agriculture Association (OAA) is founded in Albania.
1999	In Tunisia, a national regulation is issued, and the Technical Centre for Organic Agriculture (CTAB) is created. In the same year, IAMB launches the Mediterranean Organic Agriculture Network (MOAN).
2000	The organic movement in Algeria begins and an organic farming unit in Malta at the Ministry of Agriculture is established.
2001	An organic farming society in Jordan in the Ministry of Agriculture and the Ministerial organic unit in Tunisia are founded. In the same year, the first symposium on organic agriculture in the Mediterranean takes place in Morocco. Croatia and Macedonia adopt national regulations.
2002	Algeria establishes a national unit for control and certification in the Ministry of Agriculture.
2003	The organic committee in the Syrian Ministry of Agriculture is founded. In the same year, the first Arab Conference on Organic Agriculture takes place in Tunisia, followed by various other initiatives in the region, underscoring the great expansion of the sector. The Middle East Natural & Organic Products Expo is inaugurated in Dubai.
2004	An organic action plan is elaborated in Tunisia.
2005	The Association of Lebanese Organic Agriculture (ALOA)is founded.
2006	Libancert, a Lebanese Certification Body, is launched.
2007	Macedonia prepares its national action plan.

Statistical development and networking

Despite the substantial developments experienced by the sector over the last several years, access to reliable statistical data and information about the state of the art and scope of organic agriculture, particularly in the Mediterranean Basin, remains a challenge. In response to the growing need to monitor growth trends in the sector, IAMB launched the Mediterranean Organic Agriculture Network (MOAN) in 1999. Its primary objective is gathering detailed data on organic agriculture in the CIHEAM-member states (Albania, Algeria, Egypt, Greece, Italy, Lebanon, Malta, Morocco, Portugal, Spain, Tunisia and Turkey).

In 2006, in view of strengthening the Euro-Mediterranean partnership, and in consideration of the institutionalization of the organic sector by governments, especially in the non-EU states, MOAN was reorganized and now includes 24 Euro-Mediterranean countries (table 2). The Ministries of Agriculture of each member country actively participate, with representatives from the units in charge of organic agriculture. The aim of MOAN is to promote the organic sector in both the EU and non-EU Mediterranean countries and to assign an important role in representing and promoting, on the international arena, the identity and specificities of Mediterranean organic agriculture.

The member countries of the Mediterranean Organic Agriculture Network (MOAN)

European countries	Eastern Adriatic countries	Eastern and Southern Mediterranean countries
Cyprus*	Albania	Algeria
France	Bosnia& Herzegovina	Egypt
Greece*	- Federation of Bosnia and Herze-	Jordan
Italy	govina	Lebanon
Malta	- Srpska Republic[1]	Libya
Portugal*	Croatia	Morocco
Spain	Macedonia	Palestine
Slovenia	Montenegro	Syria
	Serbia	Tunisia
		Turkey

* Countries that have not yet officially joined.

In order to achieve these goals, reliable and detailed information about the sector, including its strengths and weaknesses, are needed; an objective assessment of organic agriculture's multifaceted potential and its impacts on Mediterranean farming systems and rural communities is essential.

Therefore, data collection, analysis, dissemination, and impact evaluation represent key activities of MOAN.

The second MOAN meeting held in June 2007 in Izmir, Turkey, focused on the 'Methodological issues in data collection on organic agriculture in the Mediterranean.' Nineteen countries participated, and on behalf of their ministries representatives committed to carry out the following activities:

- review the current state-of-the-art of organic data collection and reporting in MOAN countries;
- exchange best practices in organic data collection, analysis and dissemination;
- identify common problems and challenges related to organic data collection and reporting; and
- launch a long-term program of activities and initiatives for the development of a shared framework for the collection and reporting of reliable data on the Mediterranean organic sector.

Efforts are underway to prepare a joint plan of action, which will be further discussed at the next MOAN meeting to be held in Syria in March 2008.

---

[1] Republika Srpska (RS) is one of the two entities of Bosnia and Herzegovina.

151

Structural aspects

Currently, in the Mediterranean region (25 countries) there are more than five million hectares of organic land and almost 140'000 farms (see table).

Organic agriculture is now practiced in all Mediterranean countries without exception, and the sector is becoming an increasingly important market segment in all parts of the Mediterranean, albeit substantial differences in levels of development persist.

Distribution of organic land in the Mediterranean 2006 (including wild collection)

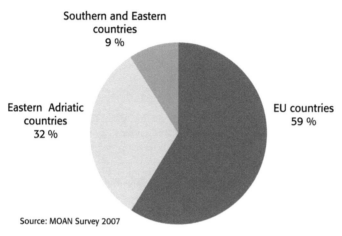

Source: MOAN Survey 2007

Figure 21: Distribution of organic land in the Mediterranean 2006 (wild collection included)

Source: MOAN Survey 2007

In general, EU Mediterranean countries have access to far greater resources and are much further developed than the Southern and the Eastern countries, as a result of their geopolitical, economic, social and legal integration. Indeed, 59 percent of the total Mediterranean organic land area is located here (see figure), the leading countries being Italy and Spain.

Table 25: Organic farming in the Mediterranean: Land area and operators 2006

Region	Subgroup	Country	Certified agricultural area (ha)	Total certified area (ha)[1]	Farms / Operators
North	European Union	Cyprus	1'979	1'979	305
		France	552'824	552'824	17'477
		Greece	302'264	302'264	24'666
		Italy	1'148'162	1'148'162	51'411
		Malta	20	20	11
		Portugal	269'374	269'374	1'660
		Slovenia	26'831	26'831	1'992
		Spain	926'390	926'390	18'318
		Sub total	3'227'844	3'227'565	115'840
	Eastern Adriatic	Albania	171	1'201	93
		Bosnia Herzegovina	714	488'804	60
		Croatia	6'012	23'670	342
		Macedonia	509	2'101	104
		Montenegro	25'051	158'851	15
		Serbia	906	1'102'906	48
		Sub total	33'363	1'777'534	662
East	Mashrek	Turkey	100'275	192'789	14'737
		Egypt	14165	14165	460
		Israel	4058	4058	263
		Jordan	1'024	1'024	25
		Lebanon	3469.56	3'470	213
		Palestine	641	641	303
		Syria	30'493	30'493	3'256
	Maghreb	Algeria	1'550	2'400	61
		Libya	n.a.	n.a.	n.a.
		Morocco	4'216	104'216	n.a.
		Tunisia	154'793	220'476	515
		Subtotal	314'685	573'732	20'191
		Total	3'575'892	5'578'830	136'617

Source: MOAN Survey 2007, with some data from the FiBL Survey
Editors' comment: Please note: In some cases, the data provided by the members of the MOAN network differ from those that were provided for the global organic survey by FiBL

A comparison of year-end data for 2004 and 2006, however, indicate particularly strong growth in countries outside the EU, such as the Eastern Adriatic countries and in countries on the Southern and Eastern shores of the Mediterranean. Among Eastern Adriatic countries, Bosnia and Herzegovina and Serbia demonstrated strongest growth and Turkey, Tunisia and Syria confirmed exceptional increases of more than 60 percent. Growth in EU countries has leveled off but remains strong, bolstered by a major increase in Spain and Portugal (see figure). In total, the organic land area increased by 2.25 million hectares, representing 70 percent more than in 2004.

---

[1] Including wild collection

As in previous years, Italy leads the top ten countries with most organic land. With an increase of 900'000 hectares, most of which is constituted by pasture and wild collection areas, Serbia has moved up the second position, followed by Spain, France and Bosnia and Herzegovina (see figure).

Increase of organic land area in the Mediterranean countries 2004-2006 (including wild collection)

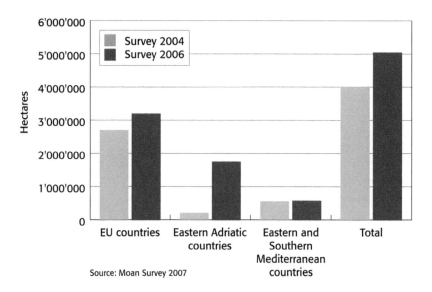

Source: Moan Survey 2007

Figure 22: Increase of organic land area in the Mediterranean 2004-2006 (wild collection included)

Source: MOAN Surveys 2005 and 2007

Examining the ten countries with the highest increase of organic land area, Serbia ranks number one, followed by Bosnia and Herzegovina, Spain, Portugal and Italy. Notably, six of the countries with the highest increase are non-EU countries, bringing the share of the non-EU countries to more than a third of the total Mediterranean organic area (as 2006, see figure). Compared to the previous year, this share expanded significantly, from 25 to 41 percent (Al Bitar, 2007), signifying sector maturation. The increase is due to diverse factors, including increase of awareness by consumers, expansion of market opportunities and government support. The increase of organic land in the area is also due to growth of the certified organic land in Serbia, Bosnia & Herzegovina and Montenegro, which is mainly wild collection and should viewed with a note of caution.

Organic farming in the Mediterranean countries - the ten countries with most organic land 2006 (including wild collection)

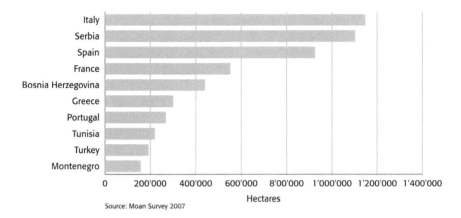

Source: Moan Survey 2007

Figure 23: The ten countries with the largest organic areas in the Mediterranean (including wild collection) 2006.

Source: MOAN Survey 2007

Legal and Policy framework

The legal framework has also evolved over the last several years, although differences between individual countries remain.

As shown in the following table, out of 25 Mediterranean countries, 17 have instituted national regulatory frameworks. Apart from the EU countries where the EU regulation 2092/91 is applied, all Eastern Adriatic countries, except the Federation of Bosnia and Herzegovina, where the regulation is still in draft form, have developed a national regulation. Among Southern and Eastern Mediterranean countries, only Israel, Tunisia and Turkey have developed their own regulations. Five other countries have regulations in draft form (Algeria, Lebanon, Egypt, Morocco and Syria), and only three entirely lack a regulatory framework.

Although certification in non-EU Mediterranean countries was performed primarily by foreign organizations until recently, nine countries now have local certification bodies, including Albania, Egypt, Lebanon, Turkey and the Eastern Adriatic countries.

Policies supporting the organic sector are still lacking in many Southern and Eastern countries. Only Tunisia, Turkey and Algeria subsidize organic farmers directly through financial support, while all Eastern Adriatic countries have already instituted support strategies.

155

Overview: Institutional framework of organic agriculture in the Mediterranean

| Shore | Country group | Country | National legislation | Support Policies | | Certifi-cation Bodies | Pro-ducers' Asso-ciations | Market | |
				Subsidy	Action Plan			Export	Local
North	Med-EU	Cyprus			draft	1 foreign; 1 local	2	null	Fair
		France				1 foreign; 5 local	78	developed	Developed
		Greece				7 local	n.a.	growing	Developed
		Italy				5 foreign 16 local	6	developed	Developed
		Malta				1 local		limited	Limited
		Portugal				2 foreign; 4 local	n.a.	n.a.	n.a.
		Slovenia				1 foreign; 1 local	10	null	Growing
		Spain				2 foreign; 25 local	36	developed	Growing
	Eastern Med.	Albania				2 foreign; 1 local	3	limited	Limited
		Bosnia and Herzegovina	Rep. Srpska; (BiH draft)			4 foreign; 1 local	n.a.	limited	
		Croatia				7 local	30 associ-ations; 15 coop.	Limited/ null	growing
		Mace-donia				3 foreign 1 local	6	null	
		Monte-negro				5 foreign; 1 local	3	limited	
		Serbia				6 foreign; 2 local	4	growing	
South and East		Turkey			draft	7 foreign; 3 local	5	developed	developed
	Mashrek	Egypt	draft			3 foreign; 2 local	8	developed	growing
		Jordan				1 foreign		limited	
		Lebanon	draft			1 local; 1 foreign with local office	2	limited	growing
		Israel			n.a.	3 local	2	developed	developed
		Palestine				1 foreign	n.a.	limited	
		Syria	draft			no	no	limited	
	Maghreb	Algeria	draft			2 foreign	2	limited	
		Libya							
		Morocco	draft			2 foreign	1	developed	
		Tunisia				4 foreign	9	developed	limited

Market aspects

Organic products are increasingly traded internationally. The market share is still modest, but trends indicate that there is an enormous potential for expansion.

Other than in Europe, the development of organic farming in the Mediterranean took place almost solely for market reasons, stimulated by the high demand of foreign markets.

Main production of the non EU Mediterranean countries grouped by geographic clusters

Cluster	Country	Main organic products
Eastern Adriatic	Albania Bosnia and Herzegovina Croatia Macedonia Montenegro Serbia	Cereals Medicinal herbs, aromatic plants Fresh vegetables Green fodder Fruits and berries Grapes Bee pastures Wild collection products Olives (Albania and Croatia)
Mashrek	Egypt Jordan Lebanon Palestine Syria Turkey	Cereals Fresh vegetables Medicinal herbs, aromatic plants Fruits Grapes Citrus Olives and olive oil
	Jordan	Date palm
	Lebanon	Wild collection products Animal products (eggs, goat milk, dairy products) Essential oils Dried fruits Processed products
	Syria	Cotton
	Turkey	Dried pulses Root crops Cotton Oilseeds Green fodder Pastures and meadows Wild collection products Dried fruits Animal products (Bovine, Ovine, Poultry, Bees) Processed products
Maghreb	Algeria Libya Morocco Tunisia	Cereals Fresh vegetables Medicinal herbs, aromatic plants Green fodder Fruits Date palm Citrus Olives Grassland Wild collection products Argan (only in Morocco)

The market differs markedly from country to country, but generally, organic products are primarily oriented towards export markets. Most of the produce (see table) is destined to foreign markets, including the EU, USA, Japan and the Gulf area. The main exports are products that either cannot be grown or have limited availability in those countries, such as spices, medicinal plants, olive oil, tropical fruits, vegetables and citrus.

Domestic markets are indeed growing, though at a indisputably sluggish pace. In reality, awareness and consciousness of the benefits of organic agriculture are on the rise, and the number of supermarkets, open markets and specialized shops are increasing.

Outlook

Despite progressive developments and access to seemingly insatiable major international markets, the evolution of the organic sector in the Mediterranean region is constrained in a number of critical ways.

Monitoring the sector's growth in the region is a complex task, due to the heterogeneity and unreliability of data and the difficulty in obtaining detailed statistical information about foreign trade, consumption and processing. In order for overcome these deficiencies, public institutions need to take responsibility for the collection and dissemination of this information, ensuring linkages between all parties engaged in the production, processing and trade of organic products.

The slow development of local markets, especially those in the Eastern and Southern Mediterranean countries, represents a significant constraint to growth and an obstacle to the long-term sustainability of the sector. Many of these local markets remain small and insufficiently organized to take advantage of the growing consumer awareness and demand. Many farmers lack information and support, and in export oriented markets, the traders play the biggest role. Moreover, due to the absence of expertise and infrastructure for the distribution and processing of organic products, raw goods are traded and exported by foreign companies, who in turn process and label the products for sale outside these countries. It is not unusual to see, for example, that raw goods are exported to European countries, only be imported back into major cities as processed products into originating countries, thus blocking further developed of processed products there and taking market share and the economic benefits from local producers.

A key goal and strategy for Mediterranean countries should be to build internationally competitive agriculture and industrial production systems, creating high quality value-added products on the market that would be both competitive with and complementary to the European brands.

In addition, many countries in the Mediterranean region contend with a lack of technical and scientific knowledge about the application of organic production methods, which includes techniques developed for areas characterized by widely varying soil, climatic and cultural conditions. Most of all, the pervasive concepts of 'commercial-organic' or organic agriculture as input substitution in the Mediterranean should be supplanted by the notion that organic agriculture is a process-based system that aims to achieve ecological, social and economic sustainability.

Organic agriculture represents a significant opportunity for farmers to diversify their production and increase income, to generate new job opportunities, to create additional value for local production, to empower women and to provide an engine for rural economic development and recovery, especially in less privileged and marginalized areas.

Finally, institutional capacity needs to be strengthened; insufficient training and expertise, lack of research and experimentation, absence of support policies and measures to further promote the organic sector constitute major weaknesses in Mediterranean countries.

Public institutions should establish programs to support and enhance production and national consumption of organic products. Supportive policies need to be developed, and efforts need to be made to establish equivalency and mutual recognition of regulations across the region and internationally.

In conclusion, organic farming represents an important opportunity for the economic development of the Mediterranean that will support vibrant rural economies. The primary task is to embed the sector's development in local culture, linking it not only to production, but also to support for local markets, social networks and environmental and health related benefits.

As envisioned by the Barcelona Declaration of the 28th November 1995, the establishment of the 'free trade' zone between the EU and the Partner countries by the year 2010 may represent the starting point for a Euro-Mediterranean shared prosperity, which may translate into an important process of evolution of the organic sector (Pierleoni and Al Bitar, 2007).

References

Al-Bitar (Ed.) (2004). Report on Organic Agriculture in the Mediterranean Area – Mediterranean Organic Agriculture Network, Options Méditerranéennes, Series B: N°50, CIHEAM- IAMB, Bari.

Al Bitar L. (2007). Organic farming in the Mediterranean region: statistics and main trends. *Organic Agriculture World Wide. Statistics and Emerging Trends 2007.* Willer H. and Yussefi Y. (2007). International Federation of Organic Agriculture Movements (IFOAM), DE-Bonn and Research Institute of Organic Agriculture (FiBL), CH-FiBL

MOAN (2006). Personal communication

MOAN (2007). Personal communication

Pierleoni D. and Al Bitar L. (2007). Italia potenziale guida del biologico mediterraneo. L'informatore Agrario, 3/2007, p. 2-4.

Country Report: Organic Food and Farming in Egypt

PAUL RYE KLEDAL[1], AHMED EL-ARABY[2] AND SHERIF G. SALEM[3]

Egypt is almost the size of Britain, France and Germany combined, yet most of its 76 million people are forced by geographical factors to live along the river Nile. Only for percent of the land in Egypt can sustain cultivation and life.

The river Nile itself, winding its way in an S-shape from the South of Egypt, on the border of Sudan, to the North running into the Mediterranean Sea in a large delta triangle, can be divided into three geographical zones: the Upper Nile, the Middle Nile and the Nile Delta. Approximately 98 percent of Egypt's population is concentrated along the Nile valley, and most in the Nile Delta, clustering around the two major cities Cairo (20 million) and Alexandria (5 million). The Nile Delta itself accounts for 30 percent of the whole Nile Valley.

Geophysically, the Nile Valley divides Egypt into two desert areas: Western Egypt, which covers more than 68 percent of the country, and Eastern Egypt, which covers 20 percent. Two percent of the population lives in oasis or reclaimed irrigated desert land.

Three types of farming can be derived from Egypt's geophysical and socioeconomic factors: farming in the oasis, Nile Valley farming and reclaimed desert land farming. The organic farms in the reclaimed desert areas are generally large and modern farms, due to various political and economic incentives promoting foreign investments and lower prices for land, whereas farms in the oasis and along the Nile Valley are small.

Egypt's approximately 460 organic farms covered an area of 14'165 hectares in 2007, and almost half of the farms were located in the Middle Nile, concentrated in the region of El-Fayoum, 100 km south of Cairo. More than half of the organic farms in Egypt are 4.5 to 20 hectares in size. There are only a few farm enterprises larger than 1000 feddan (ca. 400 hectares), but they account for 20 percent of all organic farmland, and are located in reclaimed desert land in the Nile Delta and in the Upper Nile.

Due to the favorable climate conditions and access to water in the cultivation areas, crops are harvested all year round. However, some crops are typical winter crops, like cucumbers, onions, various herbs and fodder, including alfalfa. The primary summer crops are grapes, baby corn, peas, rice and cotton. Potatoes and tomatoes are typically grown all year round.

History

Historically, the development of organic agriculture in Egypt commenced with the establishment of the biodynamic farm Sekem (hieroglyphic transcription meaning 'vitality of the

---

[1] Dr. Paul Rye Kledal (corresponding author email: paul@foi.dk) Institute of Food and Resource Economics, University of Copenhagen, Denmark
[2] Professor Ahmed El-Araby, Department of Soil Science, Ain Shams University-Cairo, Egypt
[3] Sherif G. Salem, Ph.d. student, Department of Soil Science, Ain Shams University-Cairo, Egypt

sun') in 1977 by Ibrahim Abouleish. The farm, situated on the border of desert land in the Nile Valley near the town of Belbeis in the region of Sharkia, started with 20 hectares, but expanded rapidly to 70 hectares. The first large economic venture of Sekem was the production of the medicinal compound ammoidin (a plant extract), and then manufacture of herbal teas for domestic sale and export Europe, mainly Germany. To secure supplies for a growing demand, especially for herbs and medicinal plants, Sekem leased more land and started to contract with other farmers, helping them to convert to biodynamic production methods. Similar export initiatives were taken among smallholders in the Region of El-Fayoum, which explains the large number of organic farms in this region most of which produce herbs and medicinal plants.

Organic farmer with his children picking organic hibiscus flowers for tea production

Picture: Paul Kledal, University of Copenhagen, Denmark

In 1990, Sekem founded the Centre of Organic Agriculture in Egypt (COAE) together with German and Swiss partners to serve as a local inspection body. In 1997, COAE registered as a limited liability company by Sekem, IMO (Institute of Market Ecology – Swiss based) and Demeter International for inspection and certification. Later in the same year, COAE was accredited by DAP, a German accreditation organization.

Due to dissatisfaction about pricing and contract policies among suppliers, a new certification and inspection organization was founded in 1995 - the Egyptian Centre for Organic Agriculture (ECOA). Parallel to this move, the Union of Traders and Growers of Organic and

Biodynamic Agriculture (UGEOBA) was established.

In 2007, four certification bodies were active in Egypt: COAE, ECOA, and two international bodies, the Soil Association and the Mediterranean Institute of certification (IMC). COAE and ECOA certified most of the organic farms in Egypt.

Legislation

Official legislation regulating or otherwise supporting organic agriculture has not yet been issued in Egypt. However, the draft 'Regulation to produce, process and handling organic products in Egypt, Part one' is in the process of being developed. A committee assigned by the Agriculture Council in Ministry of Trade prepared this draft. In order to become effective, this draft has to be submitted to the National Assembly for ratification. When this law is ratified, it may then be submitted to the European Union for endorsement. If endorsed and accepted by the EU, crops produced and exported from Egypt will gain access to the European markets more easily. However, many Egyptian crops are similar to those of Southern Europe, and as a result of this competition, obtaining market share entry would be expected to be a gradual process.

Market

Exports to Europe drive organic production in Egypt. The major crops exported are herbs, vegetables and fruits. The most important herbs are chamomile, coriander, dill, lemongrass, hibiscus, marjoram, parsley, peppermint and spearmint.

For vegetables, the most important crops are potatoes, onions, garlic, green beans, peppers and peas. The major fruit crops for export are various citrus, mangos, grapes and olives.

Egyptian organic exports benefit from the off-season supply to the European market, and especially potatoes and onions have found a niche due to storing and preservation problems of these products in countries with cold and humid climates.

Domestic sales of organic products are concentrated around the urban areas of Cairo and Alexandria. Outlets in high-income areas with a presence of affluent foreigners account for much of the market. Shopkeepers estimate 60 to 70 percent of the consumers are foreigners. The Sekem farm used to have fifteen of their own outlets in Cairo, but have now reduced the number to three, due to expanding sales of their products through supermarkets like Carrefour, Metro, Alfa, and Spinneys. Most of the outlets selling organic products are placed in large shopping malls. Many pharmacies in Cairo sell organic medicinal plants, essential oils and teas, and in all the types of outlets mentioned above, Sekem is the main supplier.

Package of Egyptian organic dates from SEKEM under their brand ISIS

Picture: Paul Kledal, University of Copenhagen, Denmark

Future prospects

Growth in the Egyptian organic sector aimed primarily for export could be expected to take place in the eastern part of the Nile Delta, northern part of the Sinai desert, as well as in the Upper Nile. These areas are connected to governmental land reclamation plans, and are supported by a good infrastructure linked to both harbor and airport facilities. Likewise, the tourism industry, which has become the most important contributor of foreign funds to the Egyptian economy, could be an area of future domestic growth. Organic farms in the Upper Nile are expanding through sales to hotels and restaurants in tourist places like Luxor, Hurghada and Sharm el Sheikh on the Sinai Peninsula.

However, the individual market actors face high transaction costs in searching, coordinating and establishing trustworthy partnerships, as well as in achieving the critical supply demanded by modern supermarket chains, whether domestically or for export. In this regard, small farmers often complain that they can only sell one or two of their products with premium prices. The rest, which are necessary parts of the crop rotation system, are often sold through conventional channels. Similarly, the lack of a regulation for organic agriculture in Egypt automatically creates higher costs and necessitates additional control measures for producers and exporters.

Therefore, there is an urgent need for governmental action to ratify the regulation to govern organic agriculture and to begin building institutions to support organic farmers. The government can also play a crucial role in improving market access and helping smallholders to establish marketing or producer associations, which would aid in the development of supply chains.

Websites

www.ecoa.com.eg; www.coae.com.eg; www.sekem.com

LATIN AMERICA

404'118 ha

15'443 ha

47'032 ha

1'810 ha

12'866 ha   437 ha

12'110 ha   60'000 ha   67 ha

7'469 ha

10'711 ha   15'712 ha

5'267 ha   50'713 ha   109 ha

50'473 ha

1216771 ha   880'000 ha

41'004 ha

17'705 ha

9'464 ha   2'220'489 ha

930'965 ha

Organic Area

- \< 1 %
- 1 - 5 %
- 5 - 10 %
- \> 10 %

© SOEL, Source: FiBL Survey 2008

Map 5: Organic farming in Latin America 2006

Organic Farming in Latin America

ALBERTO PIPO LERNOUD[1]

Traditional Farming

Latin America has ancient agricultural traditions; organic methods have prevailed for millennia. Rotations, variety selection, fertility management that includes composting and mulching, sophisticated irrigation systems, long-term planning and community land management were all features of American agriculture two thousand years ago. All these traditions are alive in the farmers of indigenous descent along the mountain ranges, from Mexico to Argentina. Hundreds of thousands of small farmers are now uniting through associations to rejuvenate the dignity of their knowledge within the organic movement, using the Internal Control System to certify their crops. Many of those families produce coffee, cocoa, sugar, bananas or other organic crops for export and have a small vegetable plot for food security and bartering. Others unite to reach the weekly markets around the cities, bringing their vegetables and fruits. They are striving to make a living, but organic agriculture has allowed them to plan their harvests and find a growing market for their products.

The Market

Local Markets

Some countries in Latin America have internal markets for organic products. In Brazil, for example, producer associations, including the Eco Vida network in the southern states, cooperatively market and transport their fruits and vegetables at weekly markets in larger cities or sell at open air markets or supermarkets under name of the farmer or the brand name of the association. The Fundación Maquita Cushunchic - Comercializando Como Hermanos MCCH leads a similar movement in Ecuador on a smaller scale. In Costa Rica, vegetable producers use the slogan 'From my family to your family.'

Supermarkets

Latin American supermarkets are beginning to sell organic products; dedicated space for produce can be found in Uruguay, Costa Rica, Honduras, Peru, Brazil, Argentina and to a lesser extent in other countries. In Nicaragua, the supermarket chain 'La Colonia' sells organic products, specializing in fresh vegetables. The availability of processed products locally is sparse, in part due to the difficulty of acquiring the quantities necessary for their production. Argentina boasts a wide variety of oils, flours, honey, wine, and tea on the shelves, and some supermarket chains have developed their own organic brands or have established dedicated organic sections. Some of the products that the company Sol de Acuario had on supermarket shelves in Argentina before the economic crisis, ranging from tea to breakfast

---

[1] Lernoud, Alberto Pipo, Vice President of the International Federation of Organic Agriculture Movements, Buenos Aires, Argentina.

cereals and corn flour, are now sold under the supermarket brand 'Bells Organic,' owned by the Dutch corporation Albert Hejin. In Brazil, the locally owned Zona Sul supermarket chain has promoted organic products to their customers, sampling organic products, offering rebates and initiating press and advertising campaigns. A producer cooperative recently opened 'O Supermercado Orgánico,' the first completely organic supermarket in Brasilia; the cooperative instituted a professional management and consumer satisfaction policy to ensure its success.

Specialized stores

Most Latin American countries feature specialized stores, or health food stores, which sell products from local organic farmers to an informed customer base. Such stores often serve as a central distribution point for information about local activism and organic regulations.

In Bolivia, the El Ceibo cooperative and producer association manages 8'000 hectares, mostly cocoa, nuts, quinoa, coffee and hibiscus. Irupana, a successful company there, has more than 15 stores, 12 of them in La Paz, where they sell breakfast cereals and snacks made from native crops like quinoa or amaranth. In Chile, La Ventana Orgánica and Puranatura are stores specialized in organic products. In Argentina, El Rincón Orgánico has been operating for almost 20 years, providing customers in Buenos Aires with more than 200 varieties of organic products from all over the country, receiving New Hope Communication's 'Spirit of Organic Award' in 2004 for its pioneering work.

A growing trend is the 'consumer cooperative shop.' In many secondary cities and towns, consumers come together and organize a cooperative, rent retail space and begin selling products from farmers that are members of the cooperative. This is common in Southern Brazil through the Eco Vida Network, stores are often consumer owned, permitting both lower prices and a fair share for producers.

Popular farmers' markets

Arguably the most popular form of organic trade in Latin America is the neighborhood fair or informal farmers' market. Most towns have a space, such as a park or sports arena, where producers can sell goods directly to the public on a weekly basis, which is a good opportunity for farmers to benefit from the full price without intermediaries. Local governments often support farmers' markets by providing the market infrastructure and advertising. Although the impact of these local markets may be economically insignificant, they support the livelihood of modest peasants throughout Latin America, in total representing an important percentage of the organic market. Some markets, such as the bi-weekly market in Porto Alegre, which gathers some 300 producers and thousands of consumers, have become quite successful and a model for further development in the region.

The Peruvian NGO Red Agroecologica (RAE) has developed many weekly markets in villages throughout Peru, taking advantage of traditional local trade with indigenous communities. Similar relationships with indigenous communities can be found throughout South and Central America.

In Uruguay, the Parque Rodó market began in the early 1990s, and a market has also been established in Tacuarembó. APODU, the small farmers' association that runs the markets there, has developed a Participatory Guarantee System (PGS) to ensure the authenticity of organic products, which was inspired by the successful examples in Brazil and Peru,

In the Dominican Republic, the FAMA ecological market in Santo Domingo is broadening consumer awareness of organic products. As organic farming in Mexico mushrooms, Mexico City launched the weekly organic market Tianguis Orgánicos. In Lima, Peru the Bio Feria market proved a model of success for years, and IFOAM has published a video and book describing the process of establishing an organic market based upon their experiences.

Many producer groups in Brazil, Argentina and Peru sell organic produce at the same price as conventional produce, making a political point to let consumers choose freely, rather than marketing organic products only to the rich. Some schemes have developed a sophisticated PGS system, basing their guarantee on the direct relationship between the consumer and producer, with active involvement of the community.

Box schemes and home delivery

Another important organic trade system is the box scheme. Concentrated in metropolitan areas, producers organize a planned home delivery route with boxes containing a fixed assortment of fruits and vegetables, with the occasional addition of milk products and eggs from other farmers. This form of trade has served as the starting point for many organic producer associations and specialized shops, which grew out of a successful home delivery system. In Argentina, the consumer base created over ten years through box schemes enabled producers to gain access to supermarket shelves. Uruguay is following the same pattern, and regional groups in Brazil have been reaching the public with organic produce through home delivery for almost twenty years.

Community Supported Agriculture (CSA)

Inspired by the Japanese system Teikei and the American Community Supported Agriculture (CSA), a movement is growing in some parts of Latin America: La Comunidad Sustenta a la Agricultura. A group of around forty families gets together with a farmer and makes a plan for the whole year. They decide together what to sow and develop a budget, detailing the needs of the consumers and the farmer. The consumers provide advance payment to the farmer to start that year's production; they share the risks and fix the prices. In some areas of Southern Brazil and around Lima, Peru, successful initiatives can already be found. This system is sometimes referred to as the stock market of the future, in which the consumer takes a financial risk with the potential benefit of getting good food throughout the year.

Exports

Most organic production from Latin America remains destined for export markets. From the coffee grains and bananas of Central America, to the sugar in Paraguay and the cereals and meat in Argentina, the trade of organic produce has been mostly oriented towards foreign markets. This trend is typical in the South, with poorly developed national markets and great need of cash to pay international debts. As most of the Third World countries, the American countries south of the Rio Grand sell their raw materials, to be processed into value-added products in developed countries for their national markets.

Developement of organic production in Bolivia 1995-2006

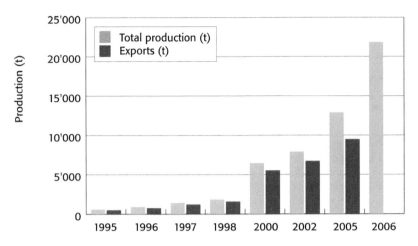

Source: Asociación de Organizaciones de Productores Ecológicos de Bolivia 2007

Figure 24: Bolivia: Organic production and exports

The graph shows the prevalence of exports, which is typical for most Latin American countries. For 2006, export data were not available.

Source: Asociación de Organizaciones de Productores Ecológicos de Bolivia

It is very difficult for small organic producers in Latin America to meet the quality standards and regulations of the demanding international markets, due to lack of information and support from governments and traders to develop quality control mechanisms.

In Costa Rica, around 30 percent of the territory is a protected natural area, and there are many organic export projects in the surrounding area, stimulated by the government. In Honduras and many other countries, multinational companies are buying land to produce organic for export. In Argentina, the well-known Italian Benetton family has bought and certified 600'000 hectares in Patagonia for organic lamb and wool production. Increasingly, European and American companies and investors' funds are buying or renting Latin American land for large-scale and technologically advanced organic production projects that benefit from relationships with buying markets and their country of origin; such projects are usually beyond the financial limitations of local companies.

Fresh fruits and vegetables

Many Latin American countries have been selling their fruit harvest to Europe and the United States. Brazil sells apples and grapes. Chile has a thriving kiwi export business, and focuses on the export of soft fruits like raspberries and strawberries. Colombia, Honduras and the Dominican Republic sell bananas, pineapples, mangos and other tropical fruits. Argentina trades apples, pears and citrus fruits, and Mexico markets apples, avocados and bananas on the world market. Seventy percent of the bananas produced in the Dominican Republic are organic. 1.7 million kg of bananas are exported annually from Costa Rica for baby food production in Europe and America. Pineapple is a growing export possibility for Central America.

Argentina, Brazil and Chile are strong vegetable exporters, both fresh and dried. In addition, Costa Rica, and other Central American countries sell smaller quantities of fresh vegetables to external markets.

Grains and cereals

Paraguay is a big organic soybean producer, together with Argentina, Mexico and Brazil, which produce and export organic corn and wheat. Organic grain farmers in several southern countries are confronting the increasing cultivation of genetically modified soy and corn.

Coffee

Mexico is the largest organic coffee producer in the world, with tens of thousands of tons of coffee beans, mostly harvested by small indigenous farmers, reaching the world's biggest supermarkets and coffee shops. Guatemala and other Central American countries have significant levels of coffee production with very similar characteristics. Coffee production is primarily defined by ecological forest management systems, creating a valuable alternative to the deforestation process that is taking place in the region.

Table 26: Organic coffee production in Latin American countries 2006

Land use details were not available for all countries.

Country	Organic coffee area (hectares)
Colombia	7'531
Cuba	490
Dominican Rep.	8'135
Ecuador	4'351
El Salvador	3'325
Guatemala	9'870
Honduras	11'732
Jamaica	2
Mexico	150'043
Panama	40
Peru	72'095
Total	267'613

Source: FiBL Survey 2008

In a recent study by Jorge Vieto, from Centro de Inteligencia Sobre Mercados Sostenibles (CIMS), there are around 63'000 organic coffee producers in Latin America, averaging 2 to 4 hectares of certified land. These small producers are responsible for 90 percent of the total production. There are more than 300 certified exporters, mainly cooperatives and farmers associations, but also some private companies. A large part of Peru's and Bolivia's coffee production is already organic. When, as in 2001, the price of the coffee is too low, farmers derive more income from their diversified production, selling tropical fruits to small processing plants. In Costa Rica this alternative is called 'Organic Integrated Farms.'

Cocoa

Most of the coffee producing countries also cultivate cocoa for chocolate, usually processed in Europe under fair trade logos and certified by European companies. It is also a very important source of income for small farmers throughout Central America and the tropical areas of South America, although to date is has proven impossible to process locally.

Table 27: Organic cocoa production in Latin American countries 2006

Land use details were not available for all countries.

Country	Organic cocoa area (hectares)
Costa Rica	2'382
Cuba (2005)	1'369
Dominican Rep.	30'624
Ecuador	20'018
Guatemala (2005)	42
Jamaica	30
Mexico (2005)	17'314
Panama (2005)	4'850
Peru	9'640
Total	86'268

Source: FiBL Survey 2008

Sugar

Brazil, Paraguay, Ecuador and Argentina are some of the sugar producers in the area. Small farmer cooperatives own or manage small sugar mills.

Table 28: Organic sugarcane production in Latin American countries 2006

Land use details were not available for all countries.

Country	Year	Organic sugarcane area (hectares)
Costa Rica	2006	324
Cuba	2005	5'662
Ecuador	2006	268
Mexico	2005	853
Paraguay	2006	16'930
Peru	2006	68
Total		24'105

Source: FiBL Survey 2008

Meats

Argentina is a large beef exporter in the region, with more than two million hectares of certified meat (beef and lamb) production. There is also a strong internal market for organic meat in Argentina. Uruguay is beginning to produce organic meat, as is Brazil. In Uruguay, 99 percent of the certified land is devoted to meat production, amounting to 70 percent of the total value of organic exports.

Wines

The amount of organic land covered by organic vines in Chile has grown from 44 hectares in 1998 to 2'443 hectares in 2006. Argentina is another country developing organic grape production for wine export, especially to Europe, the US and Japan. Some European producers are investing in Argentina's central western region to take advantage of the beneficial weather conditions and the purity the groundwater. Uruguay is also exporting some wines made from organically grown grapes.

Certification

Except for Argentina and Costa Rica, which have Third Country status in the European Union, all other Latin American producers need to be re-certified by a European company to enter the market in Europe. American or European companies certify most of the export production in Latin America in any case, as the buyer often imposes the certification. The Organic Crop Improvement Association (OCIA) and Farm Verified Organic (FVO) from USA and Naturland, BCS Oeko-Garantie and the Institute fur Marktoekologie (IMO) from Europe are very active in the region.

Some local certification bodies are very well developed, such as Argencert and Organización Internacional Agropecuaria, (OIA, Argentina), Instituto Biodinamico (Brazil), Bolicert (Bolivia) and Biolatina (Peru and others). All except for Biolatina are IFOAM Accredited certifiers. Other certification agencies in operation include Ecológica from Costa Rica, Bio Nica from Nicaragua, Maya Cert from Guatemala and CertiMex from Mexico. Chile has Certificadora Chile Orgánico (CCO) and PROA - Corporación de Promoción Agropecuaria, Uruguay has Urucert and Sociedad de Consumidores de Productos Biológicos (SCPB). Argentina has more than 12 certification agencies; apart from the aforementioned Argencert and OIA, Bio Letis (EU recognized), Food Safety, Agro Productores Organicos de Buenos Aires (APROBA), Ambiental, and Fundación Mokichi Okada (MOA) are also important.

Recently, some countries have created national laws governing organic production, including Uruguay, Chile, Paraguay and El Salvador. Bolivia has issued a decree that regulating organic production. Argentina has had a national law for many years, and its system dates back to 1992. The standards in Brazil allow PGS in local and direct markets, and other countries including Peru, Mexico and Uruguay, are developing similar systems (See also the chapter by Huber et al. on regulations).

Governmental support

No Latin American country has subsidies or economic support for organic farmers. There are however, several other forms of government support for organic farming (see also country reports).

- In many Latin American countries, organic farming is legally protected through laws.
- Costa Rica, Brazil and some others have funding for research and teaching.
- Some countries have programs for the promotion of organic farming, such as Brazil.
- Argentina, Bolivia, Chile and other countries have had official export agencies helping producers attend international fairs and print product catalogues.

In general, however, the organic movement in Latin America has grown on its own accord, with some seed funding for extension and association building by international aid agencies, especially from Germany, the Netherlands and Switzerland. International trade has been stimulated by buying companies and fair trade agencies, focusing especially on some basic products like coffee, bananas, orange juice and cocoa.

In the State of Paraná, in the south of Brazil, the big bi-national organization that runs the gigantic Itaipú dam has decided to manage the Parana river basin ecologically, generating an enormous project involving thousands of towns and villages in recycling, resource management, environmental education and organic agriculture. They call it Projeto Agua Boa (Good Water Project). Brazil has systematic support for organic agriculture in the national level, with a joint effort by the Social Development, Agriculture and Environment ministries, which understand the ecological and social advantages of organic production. A far-reaching 'agroecology development' plan is managed by the Family Agriculture Secretary, involving tens of thousands of small farmers with technical and financial support.

In Guatemala, the 'Association of Progressive Women' has been working towards agricultural diversification and agroforestry, with support from the government and the Ecological Group from Suchitepequez.

Education and extension

Latin America has a great deal of activity in education relating to organic agriculture. Many universities and agricultural organizations offer teaching courses and on farm experimental projects. Cuba had a very developed teaching and research project through the Cuban Association of Organic Agriculture ACAO, and the Brazilian Instituto Biodinamico worked systematically on farm production. Agruco and Agrecol have excelled at agricultural extension work over the years, leading to a strong support for food security and farmer knowledge, especially in the Andean region.

Some agricultural universities carry organic production courses, like the La Molina in Peru, Las Villas in Cuba and Chapingo in Mexico. In October 2004, the Catholic University of Argentina started a degree program on Organic Company Management.

The Agroecological Movement of Latin America and the Caribbean (MAELA,) an international movement linking around 80 groups in many countries, has done extension work with the small farmers throughout the region for many years, focusing especially on self sufficiency and associated skills.

The Latin American Center of Sustainable Development (CLADES), lead by Miguel Altieri and Andres Yurcevic, has built a very thorough body of knowledge and experience around agro-ecology and biodiversity issues, connecting universities (especially in the United States) with farmer groups and extension agencies, publishing authoritative studies and giving lectures in all countries. IFOAM, representing the worldwide organic movement, has been supporting and aiding the spread of organic projects in the region, bringing together different parts of the movement through international events like the Sao Paolo Scientific Conference in 1992, the Mar del Plata Scientific Conference in 1998, and the Latin American IFOAM Local Markets Conference in Buenos Aires in June 2000. Recently, producers and researchers in Latin America and the Caribbean have begun to meet annually. The first meeting took place in 2006 in Nicaragua (Ochoa et al. 2006), the second was in Guatemala in 2007.[1]

Latin America, one of the biodiversity reservoirs of the world, is just beginning to become conscious of the enormous possibilities of organic agriculture. It has the farming traditions, the fertile lands and the varied climatic zones that allow it to produce almost anything in an ecological way, aiding the much-needed greening of the planet.

Latin America: Country reports

ALBERTO PIPO LERNOUD AND MAIA LOY

The country reports are based on information from experts in the respective countries. For this edition of 'The World of Organic Agriculture,' the following reports have been revised: Argentina, Bolivia, Chile, Colombia and Uruguay. For the other country reports, the most recent data were added where available.

Argentina

Argentina had 3'192'000 certified hectares in 2000, which decreased to 2'656'000 hectares in 2006 (including wild collection). Since its peak at the beginning of the 2000, the certified area has been on the decrease, although production and exports continue to grow. Ninety-eight percent of the certified agricultural land is devoted to livestock production, which decreased by 5 percent between 2005 and 2006, whereas the cropland grew by 29 percent, and thus the loss of certified pastureland accounts for the overall decrease in hectares. The

---

[1] A detailed report by Salvador Garibay on the first meeting of organic producers and researchers in Latin America and the Caribbean was published in the 2007 edition of 'The World of Organic Agriculture' and is available at orgprints.org/10506/. The proceedings by Ochoa at al. are archived at orgprints.org/10373/

main crops being produced are cereals, sugarcane, grapes and olives for processing. Meat production still dominates, especially sheep production by big farms on the slopes of the southern states of Patagonia. Sixty-four percent of the organic land is in Patagonia, owned by only four percent of the organic farmers in the country. In the North, around a third of the farms are located in one area - the Misiones Province - mostly small farmers who are part of associations to produce sugar and mate tea. The total number of farms in Argentina 2006 was 1'486.

Ninety percent of organic production in Argentina is for export, mainly to the European Union and USA. The biggest exports are cereals and oilseeds: corn, wheat, soy, and sunflower. Fruits, including pears, apples, oranges and lemons, are also exported in large quantities. Some vegetables, especially garlic, onions, and beans are exported. Production of aromatic and medicinal plants is on the rise.

In terms of processed products, olive oil, sugar, concentrated juices, honey and wine have been successful in the European and North American import markets.

Meat exports have become the predominant export for international markets, beginning ten years ago with beef, and recently growing to include Patagonian lamb. In 2006, there were 551'217 sheep and 122'478 head of cattle certified in Argentina. Demand from Germany and the UK has lead to a rapid increase in the export of organic wool, reaching production levels of nearly 600 tons. Meat exports are continuously growing, reaching 718 tons in 2006, with the EU as its main destiny (SENASA 2007). All the products mentioned have been exported for years, many of them since 1992.

The domestic market in large cities has been growing since 1990, through home deliveries, supermarkets and specialized shops, but faced a downward trend during the economic crisis in 2001 and 2002. Some important companies disappeared from the market, including Sol de Acuario (processed foods), La Recordación (chicken) and Eco Ovo (eggs), and the presence of others diminished on the supermarket shelves. Home deliveries, with a more direct relationship with consumers, were able to survive the crisis and are now on an upward trend (El Rincón Orgánico). Some delivery services offer more than 200 different products, especially vegetables, fruits, oils, teas, breads, eggs and jams. A major company (La Serenisima) is a massive producer of organic milk, with more than 10'000 hectares and many associated farms.

Argentina was the first Third World Country to have a national regulation adapted to the European Union regulation (1992) and the first to enter the third country list. There are 12 national certifiers, some of them with a strong international presence (Argencert and OIA), and two are steadily growing (Letis and Food Safety). There is no significant activity of foreign certifiers.

Argentina organized the 12th IFOAM Scientific Conference in Mar del Plata in 1998, which to date was the biggest organic event in Latin América. Several organizations are active in the country, including the Argentinian Movement for Organic Production (MAPO), which organizes events, capacity building, research projects, conferences and meetings. There is also a new chamber of commerce for organic producers, Cámara Argentina de Productores Organicos Certificados (CAPOC), a certifiers' chamber, CACERT, and many local and regional networks.

Universities are quite active in organic issues, especially the National Buenos Aires University (UBA), the Catholic University and at Salvador University. The National Agrarian Research Institute (INTA) established a section on organic farming, which is coordinated by a former IFOAM World Board Member, Pedro Gomez. INTA also boasts the biggest organic family garden project in the world, PRO HUERTA, which counted the participation of almost one million families doing home organic farming in the 1990s. It suffered during the financial crisis, but has been back in action since 2003.

Bolivia

Bolivia had 1'069'560 certified hectares and 365'052 hectares in transition to organic in 2007. Much of this production is certified organic wild forest collection, including chestnuts or fruits. Bolivia has a long agricultural tradition, which started around 3000 years ago. The tides of the Titicaca, the highest lake in the world, was utilized to irrigate hundreds of thousands hectares around Tiwanaku, the Sacred City of the Aymaras. Bolivia is now exporting old time products like quinoa and amaranth, biodiversity that was maintained by indigenous farmers over millennia through traditional farming methods.

Development of certified land in Bolivia 2004-2006

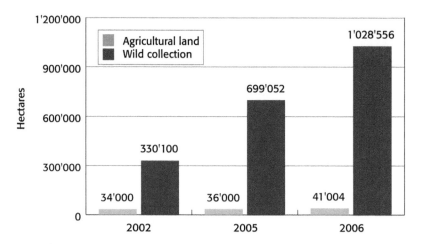

Source: Asociación de Organizaciones de Productores Ecológicos de Bolivia 2007

Figure 25: Development of certified organic land in Bolivia: Agricultural land and wild collection area

Source: Asociación de Organizaciones de Productores Ecológicos de Bolivia

The Ecological Producers Association (AOPEB) is the umbrella for 56 organizations that unite more than 30'000 producers, mostly smallholders with around one hectare.

The most important products from Bolivia are coffee, quinoa, chestnut, cocoa, vegetables, tea and herbs, and lesser volumes of amaranth, dried fruits and beans are produced. In recent years, some new products, such as banana, sesame seeds, sugarcane, soybeans and honey, have gained importance.

Bolivia has chains of shops selling organic products, especially in La Paz, Cochabamba and Santa Cruz de la Sierra. Nine Super Ecológico Markets owned by AOPEB sell only certified products, and other shops like Irupana, Eco Market, El Ceibo and Protal also sell some 'natural' products from small farmers associations.

The Chamber of Quinoa producers (CABOLQUI) actively participates at international fairs, and at BioFach 2006, they supported the IFOAM Dinner. Around 60 percent of the Bolivian organic exports of quinoa are directed to European markets and 30 or 40 percent to the US.

The government of Bolivia issued the Supreme Decree 28558 on December 22, 2005 for 'promoting the development of ecological production and setting up the National Control System.' In November 2006, Law 3525, which regulates organic production and certification, was issued. AOPEB is also developing a PGS for the local market and its own stores.

Some private institutions execute research programs on organic agriculture, including AOPEB, Productividad Biosfera Medio Ambiente (PROBIOMA) amd Promoción e Investigación de Productos Andinos (PROIMPA). Two state universities - the Agrochemical Program of the Universidad Mayor de San Simón and the Institute of Ecology - are developing organic research.

Bolivia has an IFOAM accredited national certifier, Bolicert, and many foreign certifiers acting in the country.

Brazil

In 2001, Brazil had 275'576 certified hectares. Today there are 880'000 hectares. Most of the agricultural land is permanent pasture, with cropland amounting to about one fifth of agricultural land. Additionally, 5.6 million hectares are certified for organic wild collection.

Not included in these statistics, a large number of informally certified or uncertified organic production system are part of the organic movement here. Concentrated in the southern states of Rio Grande, Parana and Sao Paulo, there is a big movement of 'family agriculture.' The number of organic producers is reportedly around 190'000, 90 percent of the farms constituted by smallholdings.

Raw organic products, including coffee, banana, soybeans, and corn constitute the bulk of exports, but the export of meat represents a rapidly growing export. A few processed products, such as concentrated fruit juices, sugar and processed soy are beginning to acquire international market share. At BioFach 2005, Brazil presented its growing organic textiles and cosmetics sector.

The domestic market in Brazil is the most developed in Latin America. Forty-five percent of the sales in the domestic market are through supermarkets, 26 percent through fairs and 16 percent in specialized stores. Most of the products are fresh vegetables and fruits, but there a growing number of processors, including both companies and small family initiatives, are marketing tea, coffee, mate tea, jams, oils, breakfast cereals and dairy products. Brasilia inaugurated the first fully organic supermarket, and an organic hamburger shop there is selling 'organic fast food.'

Twelve national and nine international certification agencies are operating in the country.

Recent developments are ensconced in an intense focus of the movement on local marketing and PGS, especially in the south, with hundreds of weekly markets. The largest among them is in Porto Alegre, with more than 300 farmers selling directly to the public every week.

There are many NGOs supporting organic farming in Brazil, mostly with small family farms. The Eco Vida Network and the Association of Organic Agriculture (AAO) are well known examples. Those NGOs, together with consumer organizations, have lobbied against the planting of GMOs, especially in southern agrarian states; in 2003, the government allowed GMOs.

Empresa Brasileira de Pesquisa Agropecuária (EMBRAPA), the national agricultural research center, is developing several research projects in collaboration with producers. The Ministry of Agrarian Development is involved promoting ecological agriculture as an alternative for the millions of small farmers throughout the country. The Ministry of Agriculture issued the 2004-2007 Plan, with six activities related to the promotion of organic farming. The 'Programa de Desarrollo de Agricultura Orgánica (Pro-Orgánico),' initiated in 2005, budgeted almost one million US Dollars[1] to stimulate the production and domestic consumption of organic products. In Brazil, organic production is regulated by the law 10.831, issued in December 2003.

Chile

At the beginning of the decade, Chile had more than 600'000 hectares, mainly Patagonian prairies dedicated to sheep production. Two major producers with more than half a million hectares, however, left the organic certification system. The current certified surface is around 39'000 hectares, of which approximately 9'000 are agricultural land; the rest is wild collection.

Chile's organic production is fully geared to exports (90 percent), and the main fresh products are lamb, apples, cherries, asparagus, blueberries, avocado, citrus, and olives. There is also a growing export of processed products, like wine, olive oil and fruit juices and concentrates. The production of organic salmon has recently begun in Chile. The export market was worth an estimated 12 million US Dollars in 2004. The main market is the US, constituting 70 percent of the total export market, followed by Europe, Japan and Canada. Today, more than 40 organic products come from 24 export companies. Concerning wine, there are 30 producers exporting, the largest with 500 certified hectares.

The internal market is very small, although there are some home delivery services in large cities (Santiago, Temuco, La Serena and Valdivia), offering mainly vegetables and fruits, and

---

[1] 1 US Dollar = 0.73096 Euros. Average exchange rate 2007.

few specialized shops (Tierra Viva, Cura Natura, La Ventana Orgánica, Pura y Natural), all of them in Santiago. In supermarkets, fruits and vegetables are present.

On January 17, 2006, the law 20089 for Organic Agriculture was issued, which serves as the national regulation for organic farming. The law allows for alternative certification systems for local markets and direct sales.

A national association as been established for the organic movement - the Agrupación de Agricultura Orgánica de Chile (AAOCH, www.agrupacionorganica.cl/).

There are three national certifiers, including Certificadora Chile Orgánico (CCO), which is reportedly the most active, the Corporación de Investigación en Agricultura Alternativa (CIAL), which also serves as an inspection background for international agencies, and the smaller Corporación de Promoción Agropecuaria (PROA). Many international agencies in this country actively support organic agriculture development projects, and some have permanent offices in Santiago. There is efficient governmental support for exports through the official agency Pro Chile. Research is carried out at the Research Institute of Chile (INIA) and the Universidad del Mar.

Colombia

The number of organic certified hectares jumped from 33'000 in 2003 to 55'072 in June 2007, covering almost 0.5 percent of the total agricultural land. The number of farms is 4'500, the majority of which are smallholders.

Development of organic agricultural land in Colombia 2002-2007

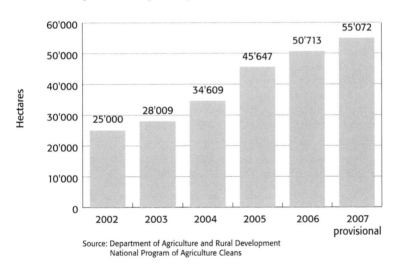

Source: Department of Agriculture and Rural Development
National Program of Agriculture Cleans

Figure 26: Development of organic agricultural land in Colombia

Source: Department of Agriculture and Rural Development. National Program of Agriculture Cleans

Forty percent of the organic land has coffee as the main crop. Colombia also produces palm oil, sugar cane, fresh and dehydrated banana, fresh mango, medicinal plants, cocoa, and some processed fruits. There is also some livestock production, including sheep and pigs.

The domestic market is very small. Some 'natural' food stores sell organic products. Supermarkets are beginning to carry some organic products, especially fruits and vegetables. New marketing initiatives work with the internet, company restaurants and schools. Some NGOs are supporting bartering schemes that include organic products.

In 1995, the Colombian Ministry of Agriculture issued the first regulation (Res. 0544/1995), which was modified in 2002 (Res. 0074). Biotrópico, a local certifier, has been in operation in Colombia since the 1990s.

The Organic Agriculture Research Center (CIAO), the National University of Colombia, the University of Santa Rosa and others have established organic production research programs.

Colombia has several groups and associations promoting organic agriculture, including the Organic Coffee Producers Association (ACOC), the Colombian Network on Biological Agriculture (RECAB), the Fundación Pro Sierra, the Corporación Penca de Sabila, the Corporación la Ceiba and Fidar. Currently, the national institute for the coordination of all organic production is in initial stages of development.

Costa Rica

In 2000, Costa Rica had 8'974 certified hectares. In 2006, that number grew to 10'711 hectares with almost 3'000 producers.

The main products for export include banana puree, cocoa, coffee, sugar, spices and medicinal herbs, blackberries, orange pulp, mango, and pineapple.

Since 1992, the farm and training center Jugar del Valle has been selling vegetables to the Mas por Menos supermarket chain. COPROALDE organized a fair in 1994, and Comercio Alternativo reached hotels, supermarkets, restaurants and schools with organic produce. Other efforts to sell organic products include ALIMCA, which focuses on home delivery, AFAORCA for coffee and APOETAR, which produces vegetables. CEDECO is an active organization that promotes research, local markets and training, and has effectively supported farmers' markets in several regions.

Costa Rica has a National Certification System that was recognized as equivalent by the European Union in 2003, and is thus part of the coveted third country list.

Since 1995, it has laws regulating pesticide use, and shortly thereafter established a national regulation for organic production in 1997, which was modified in 2000 and then again in 2001. The Inter-American Institute for Cooperation on Agriculture (IICA) has promoted a National Program of Organic Agriculture, and a law for the national promotion of organic production is under review. Costa Rica has two national certifiers - EcoLógica and the Central American Institute for the Certification of Organic Products AIMCOPOP. There are three registered international certifiers.

Mexico

In 2000, Mexico had 85'675 certified hectares. In 2002, there were an estimated 215'843 certified organic hectares. Currently, more than 400'000 hectares are under organic management. There are around 120'000 organic farms, many of which are small and owned by indigenous peoples, and use farmer group certification (Internal Control Systems). Producers here are split into two primary groups. Small-scale, low-income producers who are peasants and indigenous people that have an average of 2.25 hectares comprise one group. They associate through cooperatives, using group certification systems to decrease the cost of certification and facilitate trade. Large-scale producers, which are private enterprises that cover between 100 to 2'000 hectares and operate independently, comprise the second group. The small-scale, low-income producers constitute 98.6 percent of the total number of producers, farming 84.1 percent of the total organic acreage and generating 68.8 percent of the foreign currency earned; the remaining percentage is earned by large-scale producers. More than half of the certified land is dedicated to the production of coffee. Mexico is the world's biggest producer of organic coffee, and 80 percent of its production is coffee.

Development of organic agricultural land and export value in Mexico

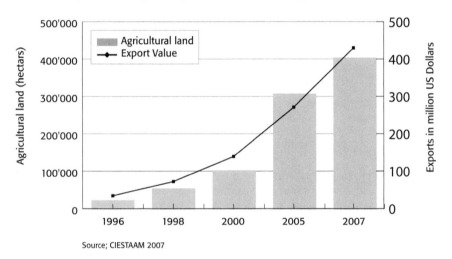

Source; CIESTAAM 2007

Figure 27: Development of organic agricultural land and of export value in Mexico

Source: CIESTAAM 2007

Most of the organic production is for export (between 80 and 85 percent), mainly to the United States and Europe. The main products are coffee, cocoa, honey, vegetables, sesame seeds, blue corn, and maguey. There is also some production of vanilla, banana, papaya, apple, avocado, medicinal plants, soybeans, oil palm and nuts. Organic exports were valued at around 430 million US Dollars[1] in 2007 The main producer states are Chiapas Oaxaca, Guerrero, Michoacán, Chihuahua, Jalisco and Veracruz.

The domestic market remains small. Only coffee and some fruits and vegetables are currently available, although there is a growing production of processed products like fruit jams and chili sauces. Herbs, honey, milk and tea are also present in some stores. Less than five percent of production sold in specialized stores in big cities (Mexico, Monterrey y Guadalajara), cafeterias, street markets and tourist areas. About ten percent of the total production that is not exported ends up being traded on the national market as conventional products. An annual organic trade fair called Exporgánicos, which aims to unite all organic producers and is promoted by the federal government, is organized in conjunction with rotating state governments. In Mexico City (Federal District) there are weekly fairs called 'Tianguis,' which are also developing PGS.

The Norm No. 37 was intended to serve as the national regulation for organic production, but is not working effectively. In November 2003, a proposed regulatory framework for organic products (Iniciativa de Ley de Productos Organicos) was presented to Mexican senators for their approval.

There are several international certifiers with offices in Mexico, including OCIA Mexico, Naturland Mexico, Bioagricert, IMO, BCS, Oregon Tilth Certified Organic, Quality Assurance International and FVO. Certimex is the most important local certification agency.

Chapingo University actively develops organic farming research, and Agricultura Ecológica Campesina (AECA) is doing on-farm research with small farmers.

The Secretary of Agriculture, in collaboration with the Bank of Mexico, has promised to finance 75 percent of certification costs in the short term.

Paraguay

Organic sugar production has provided good fortune for Paraguay, as their sugar was losing market share due to falling prices of conventional sugar. Small farmers, who represent 90 percent of the production, produce the organic sugar.

The domestic market for organic products is negligible, with some vegetables sold at markets, and some sugar and mate tea are available in supermarkets. The Organic Producers Association (APRO) and the NGO Altervida have a shop in the metropolitan area of Asunción.

The Minister of Agriculture and Arasy Organica have agreed to cooperate to aid producers in capacity building and gaining technical knowledge. Altervida started the project to highlight organic farming as an economic alternative for rural poverty in Paraguay, with the support of the Dutch organization ICCO and the European Union.

---

[1] 1 US Dollar = 0.73096 Euros. Average exchange rate 2007.

Peru

In 2001, Peru had 84'908 certified organic hectares. In 2005, it was estimated that ore than 270'00 hectares were covered by organic certification (including wild collection). More than 30'000 farmers, most of them small and indigenous, produce coffee and cocoa under with group certification.

Ninety-seven percent of the production is exported, and 94 percent of those exports are coffee and cocoa. Banana is also a growing export product. The total value of exports is estimated at around 100 million US Dollars.[1] Other exported products are quinoa, cotton, pecans, Brazil nut, onions, asparagus, sesame seeds, amaranth and tomato. Although it amounts to only three percent of the production, there is a very well organized internal market, thanks to the work of Eco Logica Peru. Weekly markets - Bio Ferias - take place in Lima and the surrounding area. Bio Canasta has initiated a home delivery service and small shops and dedicated areas in the supermarkets - Isla Ecológica are increasingly common. These channels account for an annual domestic market of around half a million US Dollars. The main products sold on the domestic market are vegetables (43 percent), fruits (41 percent), beans (9 percent) and root crops, including potatoes and sweet potatoes (7 percent).

The local certification agency Inkacert has joined together with other Latin American certifiers to form Bio Latina, which has been accepted by the EU. The certification bodies SKAL, IMO and SGS Peru have offices in Lima.

Since 1998 there has been a National Commission or Organic Production (CONAPO), which unites private sector, scientists and the government. In 2003, after a very long consensus process, the National Regulation was finalized.

The small farmers' movement is particularly active in research, concentrating on the participatory development of technologies (DPT), which is coordinated by a consortium of NGOs, including the Network of Organic Agriculture (RAE), Centro IDEAS and the Peruvian Organic Producers Association (ANPEP). A successful capacity building program has been developed through the farmer-to-farmer system. The Agrarian University of La Molina has long served as a center organic farming studies and education.

Uruguay

Uruguay currently 930'965 certified organic hectares, a stunning growth from the 1'200 hectares reported in 2000. Organic farms now cover five percent of agricultural land, with 630 organic farms that produce primarily meat, honey and vegetables.

Ninety-nine percent of production is meat for export, representing the vast majority of Uruguay's exports, which were valued at approximately eight million US Dollars[2] in 2006. Other exports include wines (with an export value of 140'000 US Dollars), honey (300'000 US Dollars), rice, milk and citrus fruits. In 2004, the estimated export volume was 2.5 million US Dollars, and the internal market 1.1 million US Dollars.

---

[1] 1 US Dollar = 0.73096 Euros. Average exchange rate 2007.
[2] 1 US Dollar = 0.73096 Euros. Average exchange rate 2007.

The domestic market is very small. The main marketing channels are supermarkets, home deliveries, farmers' markets and on-farm sales. In 2005, specialized stores began offering organic products. A weekly organic fair takes place in Montevideo, which is organized by the Small Producers' Association of Uruguay (APODU). Other regional fairs are starting to be held in secondary cities.

Organic production in Uruguay is regulated by decree No. 360/992 from the Minister of Agriculture, and was modified by the decrees 434/92, 19/93 and 194/99.

Information providers for the country reports

- Alvarado, Fernando (Peru), Balance de la Agricultura Ecológica en el Peru 1980-2003, Centro IDEAS, Red de Agricultura Ecológica del Peru. Av. Arenales, 645 Lima
- Amador, Manuel (Costa Rica), CEDECO Corporación Educativa para el Desarrollo Costarricense, Apdo. 209-1009 Fecosa. SanJosé, Internet www.cedeco.org.cr
- Augstburger, Franz (República Dominicana)
- Barrios, Lidia (Mexico), SENASICA/SAGARPA, Municipio Libre N 377, 7B Esquina Av. Cuauhtemoc, col Sta. Cruz, Atoyac 03310, Mexico D.F.,
- Darolt, Moacir (Brasil), Instituto Agronomico do Parana, IAPAR C.P., 2031 CEP 80011/970, Curitiba Parana Brasil
- Elola, Sebastian (Uruguay), Movimiento Uruguay Organico
- Escobar, Carlos, Colombia
- Fernández Araya, Claudia Andrea (Chile), AAOCH Agrupación de Agricultura Orgánica de Chile/ yUniversidad de Chile
- Fonseca, Maria Fernanda (Brasil), PESAGRO RIO. Empresa de Pesquisa Agropecuaria do Estado do Rio de Janeiro, Alameda Sao Boaventura 770, Niteroi/Rio de Janeiro/Brasil
- Garcia, Jaime E. (Costa Rica), Area de Agricultura y Ambiente, Centro de Educacion Ambiental, Universidad Estatal a Distancia, San Jose, Costa Rica
- Gomez Tovar, Laura and Manuel Angel Gomez Cruz, External Researches of the center for Economic, Social and Technological Research on World Agriculture and Agribusiness (CIESTAAM) Chapingo University, Mexico.
- Herrera Bernal, Sandra Marcela (Colombia), Cámara de Comercio de Bogotá, Carrera 9# 16 21 piso 9, Bogota, Colombia
- Iñiguez, Felipe (Mexico), EcoCuexco, Apdo. 1-1770, Guadalajara, Jalisco
- Junovich, Analía (Ecuador), Consultora Proyecto SICA, Av.Amazonas y Eloy Alfaro Ed. MAG, Quito, Ecuador
- Lernoud, Alberto Pipo (Argentina), MAPO Movimiento Argentino para la Producción Orgánica, Salguero 925 (1177), Buenos Aires, Argentina
- Meirelles, Laercio R. (Brasil), Centro Ecologico, Cxp 21 IPE/RS 95240 000
- Mendieta, Oscar (Bolivia), AOPEB Asociación de Organizaciones de Productores Ecológicos de Bolivia, Juan José Perez N° 268-A (Zona Central), 1872 La Paz, Internet www.aopeb.org

- Parra, Patricio (Chile)
- Ramos Santalla, Nelson (Bolivia), Responsable Departamento Tecnico AOPEB, Calle Landaeta N 554, Casilla postal 1872, esq. c/ Luis Crespo
- Rodríguez, Alda (Uruguay), Movimiento Uruguay Organico, E-mail muo@adinet.com.uy
- Telleria Arias, Rafael (Uruguay), Movimiento Uruguay Organico
- Ugas, Roberto (Peru), Universidad Nacional Agraria La Molina
- Zeballos, Marta (Uruguay), Asociacion Rural del Uruguay, Av. Uruguay 864. C.P 11, 100 Montevideo Uruguay, Internet www.aru.org.uy
- Zenteno Wodehouse, Virginia (Chile), Certificadora Chile Organico, Almirante Riveros 043, Providencia, Santiago

Latin America: Organic Farming Statistics

Table 29: Organic agricultural land and farms in Latin America 2006

Country	Organic land area	Share of agricultural land	Farms
Argentina	2'220'489	1.72%	1'486
Belize (2000)	1'810	1.19%	No data
Bolivia	41'004	0.11%	11'743
Brazil	880'000	0.33%	15'000
Chile	9'464	0.06%	1'000 (2005)
Colombia	50'713	0.11%	4'500
Costa Rica	10'711	0.37%	2'921
Cuba (2005)	15'443	0.23%	7'101
Dominican Rep.	47'032	1.27%	4'638
Ecuador	50'475	0.63%	137
El Salvador	7'469	0.57%	1'811
Guatemala (2005)	12'110	0.26%	2'830 (2003)
Guyana (2003)	109	0.01%	28
Honduras	12'866	0.44%	1'813
Jamaica	437	0.09%	11
Mexico	404'118	0.38%	126'000
Nicaragua	60'000	0.86%	6'600
Panama	5'267	0.24%	7
Paraguay	17'705	0.07%	3'490
Peru	121'677	0.57%	31'530
Trinidad & Tobago (2005)	67	0.05%	1
Uruguay	930'965	6.11%	630
Venezuela	15'712	0.07%	No data
Total	4'915'643	0.68%	223'277

Source: FiBL Survey 2008

Table 30: Land use in Latin America

Main land use type	Main crop category	Agricultural area (ha)
Arable land	Arable crops, no details	205'674
	Cereals	11'256
	Flowers and ornamental plants	7
	Green fodder from arable land	1'085
	Medicinal & aromatic plants	34'988
	Oilseeds	5'426
	Other arable crops	6'580
	Protein crops	458
	Root crops	36
	Seeds and seedlings	1
	Textile fibers	2'971
	Vegetables	37'974
Arable land		*306'454*

Main land use type	Main crop category	Agricultural area (ha)
Permanent crops	Citrus fruit	7'670
	Cocoa	86'268
	Coffee	267'614
	Fruit and nuts	14'020
	Grapes	2'537
	Medicinal & aromatic plants	1'595
	Olives	1'570
	Other permanent crops	1'915
	Permanent crops, no details	4
	Sugarcane	24'105
	Tea	175
	Tropical fruit and nuts	87'220
Permanent crops		*494'692*
Cropland, other/no details		*58'527*
Permanent grassland		*3'792'234*
Other		*3'900*
No information		*259'835*
Total		4'915'643

Source: FiBL Survey 2008
Please note: Detailed crop statistics were not available for all countries

Table 31: Certified organic wild collection in Latin America 2006

Country	Certified wild collection area (ha)
Argentina	606'974
Bolivia	1'028'556
Brazil	5'600'000
Chile	38'578
Ecuador	5'416
Mexico	12'647
Peru	148'942
Total	7'441'112

Source: FiBL Survey 2008

Data and information sources

- Argentina: Servicio Nacional de Sanidad y Calidad Agroalimentaria SENASA (2007): Situación de la Producción Orgánica en la Argentina durante el año 2006. Buenos Aires. Available at http://www.alimentosargentinos.gov.ar/0-3/organico/informes/ORGANICOS_SENASA_2006.pdf
- Belize: Data provided by / Source: Jaime E. Garcia, Universidad Estatal a Distancia (UNED), San Pedro de Montes de Oca, Costa Rica
- Bolivia: Data provided by Nelson C. Ramos Santalla, Asociación de Organizaciones de Productores Ecológicos de Bolivia, La Paz, Bolivia, www.aopeb.org

187

- Brazil: Totals provided by Lucimar Santiago de Abreu, Sociologia - Embrapa Meio Ambiente, Jaguariúna, São Paulo, Brasil; Source: Ministry of Agriculture, MAP. Landuse data (as of 2005) provided by Alexandre Harkaly, Instituto Biodinâmico, Botucatu, Brazil; Data source: Ministry of Agriculture, Brasilia, Brazil
- Chile: Source/ Data provided by Pilar M. Eguillor Recabarren, Oficina de Estudios y Políticas Agrarias (OPEDA), Ministerio de Agricultura Teatinos 40, piso 8. Santiago, Chile and by Virginia Zenteno, Agricultura Orgánica, Santa Isabel. Source: OPEDA, based on information of the certifiers. Farms: Data from 2005.
- Colombia: Data provided by Carlos Escobar, Conexion Ecologica, and Ricardo Zorro, Camera de Comercia Bogota; Source: Department of Agriculture and Rural Development. National program of Agriculture Cleans
- Costa Rica: Source: Ministerio de Agricultura y Ganadería, Estatisticas 2006 Agricoltura Organica, http://www.protecnet.go.cr/SFE/organica1/Or_estadisticas.html download of September 17, 2007
- Cuba: Data for 2005. Data provided by: Lukas Kilcher, FiBL, Frick, Switzerland; Data source: CAI Balino, Ministry of Agriculture, Havana, Cuba
- Dominican Rep. Data provided by/Source: Franz Augstburger, IMO Caribe, Santo Domingo, República Dominicana; additional certifier data
- Ecuador: Data provided by Sonja Lehmann, GTZ Ecuador, Quito, Ecuador. Source: GTZ Ecuador and Servicio de Sanidad Agropecuaria - SESA, Quito, Ecuador. The number of farms actually refers to associations of organic farms. The total number of individual producers ist a lot higher.
- El Salvador:Data provided by: Beatriz Alegria, CLUSA El Salvador, San Salvador, El Salvador. Source: Agencia para el Desarrollo de Austria/Horizont 3000/CLUSA/CORDES/IICA (2006): Diagnóstico de Situación de la Producción Orgánica en El Salvador y una Propuesta para su Fomento
- Guatemala: Data as of 2005, provcided by Eduardo Calderón, Agexport, Guatemala
- Guyana. Data from 2003. Data provided by / Source: Jaime Castro Mendivil, Control Union, Lima, Peru
- Honduras: Data provided by Sandra Elvir Sanchez, Coordinadora de Agricultura Organica del Ministerio de Agricultura y Ganaderia, Honduras based on data by the operators
- Jamaica: Data provided by / Source: Dwight Robinson, Jamaican Organic Agriculture Movement (JOAM), Kingston, Jamaica
- Mexico: Data provided by Laura Gomez, Universidad Autónoma Chapingo, Source: CIESTAAM 2007. Landuse data as of 2005, Source: Universidad Autónoma Chapingo, Centro de Investigaciones, Económicas, Sociales y Tecnológicas de la Agroindustria y la Agricultura Mundial (PIAI-CIESTAAM) (2005) Agricultura, Apicultura y Ganadería Orgánicas de México – 2005. Situación – Retos – Tendencias. Chapingo, México
- Nicaragua: Data provided by Miguel Altamirano, Instituto Interamericano de Cooperación para la Agricultura (IICA), Managua, Nicaragua, Source: Ministerio de Agricultura y Forestal MAGFOR
- Panama: The figure supplied originally was updated with certifier data, but will probably need to be revised.

- Paraguay: Data provided by / Source: Dr. Fernando Rios, Servicio Nacional de Calidad y Sanidad Vegetal y de Semillas (SENAVE)
- Peru: Data provided by Julia Salazar Suarez, SENASA, Peru
- Trinidad & Tobago: Data provided by / Source: Gia Gaspard Taylor, Network of Non Governmental Organizations Trinidad & Tobago, Trinidad & Tobago
- Uruguay: Data provided by: Betty Mandl, D.G.S.A., Source: I.N.I.A
- Venezuela: Certifier data

Further reading

Many South and Central American countries have made detailed reports about organic farming available, including detailed statistics. Please find a list below:

Agencia para el Desarrollo de Austria/Horizont 3000/CLUSA/CORDES/IICA (2006): Diagnóstico de Situación de la Producción Orgánica en El Salvador y una Propuesta para su Fomento. Available at http://www.elsalvadororganico.com.sv/docs/Situacion_Agricult_Organica_ES_2007.pdf

Cussianovich, P. And M. Altamirano (Eds.) (2005): Estrategia nacional para el fomento de la producción orgánica en Nicaragua. 'Una propuesta participativa de los actores del movimiento orgánico nicaragüense.' MAGFOR, INTA, IICA, COSUDE, Embajada de Austria. Managua, Nicaragua. Available at http://www.iica.int.ni/Estudios_PDF/Estra_Prod_Org_Nic.pdf

EMG Consultores (2007): Estudio del mercado chileno de agricultura organic. Santiago. http://www.inforganic.com/node/1483

Fernández Montoya, Marco Vinicio ; Gregorio Varela Ochoa; Salvador V. Garibay; Gilles Weidmann (editores) (2007): 2ndo Encuentro latinoamericano y del caribe de productoras, productores experimentadores y de investi-gadores en agricultura orgánica. Memorias de resúmenes. 1 al 5 de Octubre del 2007 en Antigua Guate-mala, Guatemala. Facultad de Agronomía de la Universidad de San Carlos, Guatemala, Universidad Na-cional Agraria (UNA) de Nicaragua y el Instituto deInvestigaciones para la Agricultura Orgánica (FiBL), Suiza.

Garibay, Salvador and Zamora, Eduardo (2003) 'Producción Orgánica en Nicaragua.' Research Insitute of Organic Agriculture (FiBL), CH-Frick. Available at www.orgprints.org/2691/

Martinez, Javier, R (2007): Organic Cocoa from Peru. Promperu Commercial Report of Organic Products. Lima.

Ministério da Agricultura, Pecuária e Abastecimento – MAPA, Secretaria de Política Agrícola – SPA, Instituto In-teramericano de Cooperação para a Agricultura – IICA (2007) Cadeia Produtiva de Produtos Orgânicos. Brasília. Available at http://webiica.iica.ac.cr/bibliotecas/repiica/B0590p/B0590p.pdf

Ochoa, Gregorio Varela; Garibay, Salvador and Weidmann, Gilles, Eds. (2006) 1er Encuentro latinoamericano y del caribe de productoras y productores experimentadores y de investigadores en agricultura orgánica, 26 al 29 de septiembre de 2006, Managua, Nicaragua. Memorias de resúmenes [First Latin American and Car-ribbean Meeting of Organic Producers and Researchers, held September 26-28, 2006 in Managua, Nic-caragua. Summary of the Presentations.]. Proceedings of 1er Encuentro latinoamericano y del caribe de productoras y productores experimentadores y de investigadores en agricultura orgánica, Managua, Nic-caragua, September 26-28, 2006. Research Institute of Organic Agriculture (FiBL), Frick, Switzerland. Available at http://orgprints.org/10373/

República de Colombia, Ministerio de Agricultura y Desarrollo Rural (2007): Agricultura ecológica en Colombia. Bogotá. Available at www.minagricultura.gov.co/archivos/articulo_de_agricultura_ecologica._madr._2007.pdf

Servicio Nacional de Sanidad y Calidad Agroalimentaria SENASA (2007): Situación de la Producción Orgánica en la Argentina durante el año 2006. Buenos Aires. Available at http://www.alimentosargentinos.gov.ar/0-3/organico/informes/ORGANICOS_SENASA_2006.pdf

Universidad Autónoma Chapingo, Centro de Investigaciones, Económicas, Sociales y Tecnológicas de la Agroindustria y la Agricultura Mundial (PIAI-CIESTAAM) (2005) Agricultura, Apicultura y Ganadería Orgánicas de México – 2005. Situación – Retos – Tendencias. Chapingo, México

NORTH AMERICA

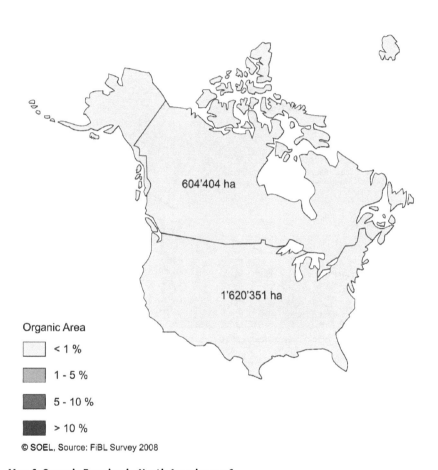

604'404 ha

1'620'351 ha

Organic Area

	< 1 %
	1 - 5 %
	5 - 10 %
	> 10 %

© SOEL, Source: FiBL Survey 2008

Map 6: Organic Farming in North America 2006

United States

Barbara Haumann[1]

The US market for organic products continues to grow at a fast pace.

According to The Organic Trade Association's (OTA's) 2007 Manufacturer Survey, US retail sales of organic products grew overall by 21 percent in 2006, to reach nearly 17.7 billion US Dollars.[2] Organic sales are projected to continue to experience double-digit growth in the foreseeable future.

US sales of organic food and beverages grew by 20.9 percent during 2006 to reach slightly more than 16.7 billion US Dollars, up from 13.8 billion US Dollars in 2005. As a result, organic food and beverage sales in 2006 represented approximately 2.8 percent of all US retail sales of food and beverages. Sales of organic food and beverages were projected to reach 20 billion US Dollars in 2007.

Organic food sales in the US

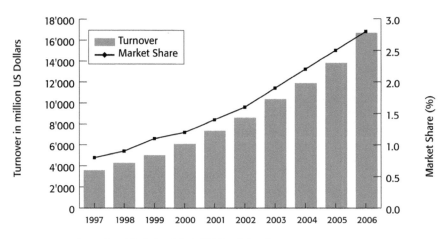

Source: Organic Trade Association's 2007 Manufacturer Survey

Figure 28: Growth of the US market for organic food 1997-2006

Source: OTA Organic Manufacturers' Survey 2007[3]

Meanwhile, US sales of nonfood organic products grew from 744 million US Dollars in 2005

---

[1] Barbara Haumann, Organic Trade Association, PO Box 547, Greenfield MA 01302, Shipping: 60 Wells Street, Greenfield MA 01301, E-mail info@ota.com, www.ota.com/
[2] 1 US Dollar = 0.79703 Euros. Average exchange rate 2006.
[3] Information on the survey is available at www.ota.com/pics/documents/2007ExecutiveSummary.pdf

to 938 million US Dollars in 2006, reflecting a 26 percent increase. Nonfood categories include organic supplements, personal care products, household products and cleaners, pet food, flowers, and fiber products such as linens and clothing.

Looking forward, the survey anticipated growth of approximately 18 percent each year on average for 2007 through 2010 for organic food products. Nonfood product sales are projected to grow anywhere from 16 percent (organic flowers) to 40 percent (organic fiber and clothing) each year on average during this same period.

Such heightened interest in organic continues to drive demand for raw materials. In the OTA survey, 55 percent of respondents reported that a lack of dependable supply of organic raw materials has restricted their company from generating more sales of organic products - up from 52 percent in the 2006 survey. This highlights the continuing need for additional measures to increase the supply of organic ingredients.

Market for organic products

Organic food categories experiencing the greatest growth during 2006 included meat (29 percent), dairy products (25 percent), fruits and vegetables (24 percent), and bread and grain products (23 percent). The fastest-growing non-food categories were organic pet food (nearly 37 percent), household products and cleaners (nearly 32 percent), fiber linens and clothing (nearly 27 percent), and supplements (26 percent).

Organic foods now compete head-to-head with non-organic foods in many venues. Mainstream grocery stores and supermarkets, representing the single largest channel for organic foods, accounted for 38 percent of organic food sales in 2006. Another 8 percent of sales occurred through mass merchandisers and club stores. Thus, more and more US consumers, regardless of income, are finding organic foods where they shop. In addition, foodservice venues offered more organic products that ever before, representing over 4 percent of total organic food sales.

Large natural food chains, along with small natural food chains or independent natural groceries and health food stores, represented nearly 44 percent of organic food sales, down from over 47 percent of sales in 2005. The biggest development on this front during 2007 was the purchase of Wild Oats Markets by Whole Food Market.

Meanwhile, 59 percent of all survey respondents said they display the USDA (US Department of Agriculture) Organic seal on their products (Note: products must contain at least 95 percent organic ingredients by weight, excluding water and salt, to be eligible for the seal on the label). Of the 41 percent not currently using the seal, 42 percent intend to use the USDA Organic seal in the future. In addition, 56 percent of respondents reported that the USDA labeling and certification programs had somewhat or dramatically increased their sales of organic products.

Although the United States imports a variety of organic products and ingredients to help meet US consumer demand, there still are no statistics available on how much or a breakdown according to category. This is because the US government does not track statistics on US imports or exports according to organic or non-organic status.

US imports of organic products and ingredients come from around the world, including Central and South America, Canada, Asia, Australia, Europe, and New Zealand. As is true for domestically produced products, in order to be sold as organic in the United States, imported products must be certified by an agency accredited by the US Department of Agriculture (USDA) as meeting US national organic standards.

Production statistics

The latest statistics available for US organic production are those released in December 2006 by USDA's Economic Research Service (ERS), which showed there were at least 8'445 certified organic farms in the United States in 2005, up from 8'035 certified organic farms in 2003. The 2005 operations represented slightly more than 4 million acres[1] under organic management, up from 3 million acres in 2004 and nearly 2.2 million acres in 2003. For the first time, all 50 US states had some certified organic farmland.[2]

ERS data for 2005 showed 1'722'565 acres of organic cropland (about 0.51 percent of all US cropland) and an additional 2'281'408 acres in pasture and rangeland (about 0.5 percent of all US pasture). Organic cropland in 2005 was up from 1'451'601 acres in 2003, while organic pasture grew substantially from the 745'273 acres recorded for 2003.

Growth of the organic agricultural land and farms in the US 1995-2005

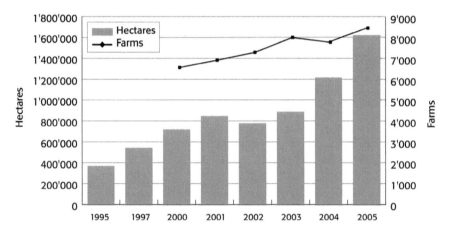

Source: USDA Economic Research Service 2007

Figure 29: Growth of the organic agricultural land and farms in the US 1995-2005

Source: USDA Economic Research Service

---

[1] 1 acre = 0.4047 hectares
[2] See U.S. Department of Agriculture's Economic Research Service: Briefing Rooms / Organic Agriculture at www.ers.usda.gov/Briefing/Organic/

Livestock numbers in 2005 were up substantially from 2003, reflecting the growing demand for organic milk and meat in the United States. The number of organically raised milk cows grew from 74'435 in 2003 to 86'032 in 2005. The number of organic beef cattle grew from 27'285 in 2003 to 70'219 in 2005. In addition, the number of organic hogs and pigs grew from 6'564 in 2003 to 10'018 in 2005. Total livestock (which included young stock and sheep) was up to 229'788 in 2005, from 124'346 in 2003. Total organic poultry—including layer hens, broilers and turkeys—reached 13'373, 270 in 2005, from 8'780'152 in 2003. According to ERS, nearly one percent of dairy cows and 0.6 percent of layer hens in the United States in 2005 were managed using certified organic practices.

Paul Tukey, founder of SafeLawns.org, is shown digging up soil on the National Mall in Washington, D.C. To show that organic lawn care techniques are effective, Safe-Lawns.org has undertaken a project to convert a section of lawn of more than four acres to organic practices at the National Mall (Picture: OTA).

ERS has indicated it hopes to have updated figures available during the first quarter of 2008. Although acreage statistics for 2006 and 2007 are not yet available, ERS sees indications that the US organic farm sector continues to grow. The number of certified organic farmers went up in over 30 states in the United States between 2005 and 2006, as existing organic farmers expanded their operations and new organic growers became certified. Markets for organic vegetables have been developing for decades in the United States, and nearly 5 percent of the US vegetable crop was produced under certified organic farming systems in 2005—a higher adoption level than for most other crops. The largest certifier of vegetable crops in the United States is California Certified Organic Farmers, which certified 11 percent more organic vegetable acreage in 2006 than in the previous year. The number of organic dairies has also increased in the past two years, and certified organic pasture is continuing to expand.

Meanwhile, ERS in late 2007 posted new information on procurement practices and use of contracts by certified organic handlers (packers, shippers, manufacturers, processors, brokers, and distributors). The interactive data contain select results from the 2004 Nationwide Survey of Organic Manufacturers, Processors, and Distributors.

Procurement information includes basic characteristics of organic handlers, their purchasing practices, and their relationship with suppliers, including important supplier attributes. Contracting information includes the use of written and verbal contracts between organic handlers and their suppliers.[1]

Recognizing the need to encourage additional farmers to switch to organic production, the Organic Trade Association during 2007 unveiled a new web site to provide a clearinghouse of resources for producers and processors transitioning to organic or interested in creating new organic enterprises. The web site, www.HowToGoOrganic™.com, features two 'Pathways for Organic'—one for farmers and one for processors—as well as a regional directory for the United States, and a searchable North American organic directory. Resource listings in the North American directory can be searched by topic and subtopic, by type of resource, or by state. The site also features profiles of farmers and businesses that have successfully become certified organic or that are working through the process.

As of December 4, 2007, there were 95 agencies accredited by the US Department of Agriculture to certify farms, processing and handling operations as meeting national organic standards. Of those, 55 were based in the United States, and the remaining 40 certifying agencies were from other parts of the world.

Standards update

Standards work during 2007 centered around a number of issues, including US organic dairy operations and the need to spell out more clearly such practices as pasture requirements and the process for converting a dairy herd to gain organic certification and to supply replacement animals. Although the National Organic Program (NOP) during 2006 issued an advanced notice of proposed rulemaking for pasture requirements, this issue was still unresolved at the end of 2007.

Other issues still being worked on by NOP and its advisory board, the National Organic Standards Board, include:

- Reviewing materials coming under the five-year sunset rule, which requires all materials listed on the National List of Accepted and Prohibited Materials to be reviewed after five years in order to remain on the list or be removed
- Reviewing 606 petitions and rule changes covering materials, including refining the definition of 'non-agricultural' substances.

---

[1] See www.ers.usda.gov/Data/OrganicHandlers/.

- Renewing accreditation of certifying agents.
- Publishing guidance on commercial availability, multi-site operation certification and inspection issues, and identifying certifiers of final handlers on labels.

Also up for discussion during 2007 was a US Department of Agriculture (USDA) interim final rule allowing the possible use of 38 non-organic agricultural ingredients in products bearing the 'Organic' label if an organic version of those ingredients were unavailable in the desired form, quality or quantity.

Prior to June 9, 2007, any non-organic agricultural ingredient, if an organic version were unavailable, could be used in the remaining five percent of ingredients in products labeled as 'Organic' (in which at least 95 percent of the ingredients by weight, excluding water and salt, must be organic). However, as the result of a court decision, only agricultural ingredients listed in Section 606 could be used in this manner after June 9.

Before the interim final rule was published, there were only five such ingredients on the list. The interim final rule, published July 27, included 38 other items.

Because of great confusion over what the interim rule meant, OTA mounted a letter-writing campaign to show that the organic system still needs this flexibility to stay healthy. For instance, the interim final rule does not provide automatic use of the 38 non-organic agricultural ingredients, but rather only allows their use if they aren't available in organic form and if this is verified by a USDA-accredited certifying agency.

Because USDA's National Organic Program is understaffed and has many tasks to manage, it is unclear how long it will take for publication of a final rule on this issue.

Farm Bill

To help further domestic organic production, OTA and other industry and farm groups during 2007 continued to encourage the US Congress to strengthen and support organic agriculture by incorporating provisions for this burgeoning sector in the 2007 Farm Bill. Currently, the sector receives only a tiny fraction of USDA's budget.

In its proposals, OTA has emphasized four priorities:

- fostering conversion to organic agriculture and trade
- eliminating hurdles to organic agriculture and trade
- initiating and funding organic agriculture and economic research, and
- maintaining and enhancing current agency programs.

In late July, the US House of Representatives approved more than 300 million US Dollars for organic agriculture as part of its version of the 2007 Farm Bill. In addition, the House of Representatives directed the Federal Crop Insurance Corporation to provide equitable crop insurance to organic producers. Currently, organic producers pay a 5 percent surcharge, and if losses occur, they are paid at the conventional, not the organic, price.

Key organic provisions in the House of Representatives' version include:

- 50 million US Dollars authorized for organic conversion, and technical and educational assistance;

- 22 million US Dollars in mandatory funding for National Organic Certification Cost Share Program;
- 3 million US Dollars in mandatory funding for Organic Production and Market Data;
- 125 million US Dollars for organic research (25 million US Dollars authorized per year, FY[1] 2009 - 2012, plus mandatory funding of 25 US Dollars million for FY 2008 - 2012);
- 100 million US Dollars authorized for urban organic gardens and greenhouses operated by residents in the neighborhood, including cooperatives, to produce fruits and vegetables and sell them to local grocery stores; and
- 5 million US Dollars in mandatory funding (1 million US Dollars per year) in the Agricultural Management Assistance Program for organic certification cost share payments to producers in 17 states.

On October 24, the Senate Agriculture Committee approved its version, which also included funding and direction for key organic priorities such as funding for organic research, data collection, and transition to organic production. In addition, the committee took steps to eliminate the crop insurance premium for organic producers.

The Senate Agriculture Committee's version of the Farm Bill:

- recognizes that increased funding is essential for the National Organic Program at the US Department of Agriculture at the full authorized level;
- includes 5 million US Dollars for organic data collection to help provide better price and yield information for organically grown crops;
- includes 22 million US Dollars in new funds to help farmers transition into organic production;
- bars USDA from charging a premium surcharge on organic crop insurance, unless validated by loss history on a crop-by-crop basis;
- adds organic production as an eligible activity in the Environmental Quality Incentives Program;
- adds to the Soil and Water Conservation Protection Loans a priority for those converting to organic farming practices and adds conversion to organic production as an eligible loan purpose; and
- provides 80 million US Dollars over the life of the bill for organic agriculture research and extension.

On December 14 in a 79-to-14 vote, the US Senate approved its version of the Farm Bill. During 2008, the House and Senate will go to conference to resolve differences between the two versions before a final bill is submitted to the White House for approval. Questions remain on whether the Bush Administration will carry out a possible veto of the final legislation.

All Things Organic™

Held for the seventh year in a row, OTA's All Things Organic™ Conference and Trade Show in May 2007 was co-located with four™ other shows (Food Marketing Institute Show, the

---

[1] FY = Fiscal year

Fancy Food Show, United Produce Expo & Conference, and the US Food Export Showcase) in Chicago, IL. Reflecting the excitement it generated, the show has twice been named among Trade Show Week's Top 50 fastest-growing shows in North America.

In 2007, the event drew 16'000 attendees, including buyers and other representatives of the mainstream and natural food industry. Much of the excitement on the show floor focused on the increased mainstream popularity of organic products, as well as an Organic for Kids product showcase, The Corner Store (a demonstration store displaying non-food organic products), and an Organic Spirits Pavilion.

The trade show itself, produced in partnership with Diversified Business Communications, drew over 10'500 attendees and featured a diverse range of products in approximately 540 booths. The accompanying conference featured a shared keynote speaker as well as two keynote speakers for All Things Organic™, five Organic 101 conference sessions, 18 conference topics, welcome party, dinner and awards ceremony, and much more.

Then U.S. Secretary of Agriculture Mike Johanns (right) speaks with Mark Bradley and Valerie Frances of the U.S. Department of Agriculture's National Organic Program at All Things Organic™ Conference and Trade Show in May 2007 (Picture: OTA)

Plans are now underway for OTA's 2008 All Things Organic™ Conference and Trade Show, slated for April 26-29 at McCormick Place in Chicago. This year, All Things Organic™ will be part of the Global Food & Style Expo along with the National Association for Specialty Food Trade's Fancy Food Show™ and the National Association of State Departments of Agriculture's US Food Export Showcase.

References

The Organic Trade Association's 2007 Manufacturer Survey, October 2007, researched and produced for the Organic Trade Association under contract by Packaged Facts.

Catherine Greene, U.S. Department of Agriculture's Economic Research Service, 'U.S. Organic Agriculture in the U.S., 1992-2005,' December 2006, www.ers.usda.gov/Data/Organic/index.htm.

Catherine Green, U.S. Department of Agriculture's Economic Research Service, personal correspondence Dec. 13, 2007.

USDA's Economic Research Service, data from the 2004 Nationwide Survey of Organic Manufacturers, Processors, and Distributors, www.ers.usda.gov/Data/OrganicHandlers.

U.S. Department of Agriculture's National Organic Program web site www.ams.usda.gov/nop

Canada

MATTHEW HOLMES[1]

Market for organic products

The Canadian market for organic products continues to grow at a fast pace. The majority of Canada's organic products are imported (estimated at more than 80 percent of organic products), with most coming from the United States. Other key source countries include the European Union, Turkey, China, Brazil, Argentina, Mexico, Indonesia, Paraguay and India.

A recent study commissioned by the Organic Agriculture Centre of Canada shows that retail sales of certified organic food in Canada were worth more than 1 billion Canadian Dollars[2] in 2006, with annual growth of over 20 percent. Nationwide, retail sales expanded by 28 percent from 2005 to 2006. Mainstream supermarket chains have responded to consumer demand and now sell over 40 percent of all organic food sold in Canada, estimated at 411.6 million Canadian Dollars in 2006 (representing close to one percent of total retail food sales). Large natural food store chains and independent health food stores accounted for another 329 million Canadian Dollars, and an estimated 174.7 million Canadian Dollars was sold though smaller grocery stores, warehouse clubs, drug stores and other specialty stores. Direct sales of certified organic produce at farmers' markets across the country and at the farm gate were estimated to be worth at least 50 million Canadian Dollars, while organic food box delivery programs brought in another 20 million Canadian Dollars. Consumers in British Columbia buy more than those in other provinces: 13 percent of the country's population buys 26 percent of organic food. Alberta showed the largest annual growth in retail sales, with a 44 percent increase, followed by British Columbia and the Maritimes (34 percent), Ontario (24 percent) and Quebec (21 percent).

Continued growth is expected over the next few years with the introduction of the 'Biologique Canada Organic' seal when the new federal Organic Products Regulations are fully implemented. This seal will be optional for all organic food products, including imported products (which must carry country of origin information in close proximity).

Production statistics

Organic farming in Canada has enjoyed a strong producer movement for over 25 years, with overall growth in organic acreage and market share across the country. The most recent statistics available from the Canadian Organic Growers (COG) show that in 2006 Canada had 3571 certified organic farmers (representing 1.5 percent of all farmers). The concentration is in field crops, vegetables, livestock and maple syrup. Acreage in organic production was 557'339 hectares, with 434'153 hectares in additional (wild) lands, and 47'065 hectares

---

[1] Matthew Holmes, Managing Director, Organic Trade Association, Canadian Office, PO Box 6364, Sackville, NB E4L 1G6, Canada, www.ota.com. Stephanie Wells, OTA's liaison in Canada, supplied assistance for this report.
[2] 1 Canadian Dollar (CAD) = 0.68215 Euros = 0.93519 US Dollars. Average exchange rate 2007.

in transition. In 2005, Canada also had 817 certified organic processors and handlers.[1]

Countdown to implementation

Although Canada has had a strong organic standard since 1999, it had been voluntary and not supported by regulation. This will change in late 2008, when Canada's Organic Products Regulations (OPR) are fully implemented by December 14. These regulations will make the Canadian Organic Standards and Permitted Substances List (PSL) mandatory for all organic food and livestock feed products sold in Canada, regardless of organic status under other regulatory programs. All products for sale in Canada will have to be certified to the Canadian standards and accredited by an accreditation body recognized by the Canada Organic Office (COO). OTA staff in Canada have encouraged Canadian officials to engage in equivalency discussions with Canada's two major organic markets—the United States and EU. However, in the meantime, OTA has encouraged all who have business in Canada to speak with their certifying bodies and suppliers regarding certification to the Canadian standards.

At the end of 2007, the regulations were undergoing an amendment process, mostly based on issues flagged by OTA and others in the industry, regarding the regulations' scope, import requirements, labeling requirements, roles and responsibilities of certifying bodies, accreditation bodies and the Canada Organic Office (COO) of the Canadian Food Inspection Agency. It is expected that the amended OPR will be re-published in early winter 2008 for public comment.

In Canada, the organic standards are published separately from the regulations, and are directly controlled by the organic sector. The Canadian General Standards Board's (CGSB's) Committee on Organic Agriculture, a committee of industry representatives, is fine-tuning the Canadian Organic Standards and Permitted Substances List (PSL) to meet legal requirements, harmonize with the proposed national regulations, and clarify certain guidelines within the standards. The committee held meetings in August and November 2007 in Ottawa to discuss these issues, and has scheduled its last set of meetings on this current round of standards housekeeping for April 2008. This committee has also endorsed OTA's recommendation that it begin discussing specific processing guidelines for organic non-food items such as textiles and personal care products. The Canadian Organic Growers, which is providing logistics planning and administration for the process with the Canadian General Standards Board, secured the government support to update the Canadian standard.

Canada's Permitted Substances List (PSL) is a 'positive' list of those substances permitted in organic production. Only substances included on the list may be used in organic production. As a result, companies that grow or process organic products for the Canadian market, no matter where they are based, are encouraged to examine the list to be certain that all substances used in formulations are included. If not, they must be submitted to CGSB to be reviewed for possible inclusion. If a company uses a substance not on the PSL, it risks losing the ability to market that product as organic in Canada as of December 2008.

---

[1] The statistical reports of the Canadian Organic Growers are available at www.cog.ca/OrganicStatistics.htm

Industry developments

There has been no official announcement of an expansion of Canada's list of Harmonization Serial (HS) Codes announced last year. However, Agriculture and Agri-Food Canada with Statistics Canada have been consulting with OTA in Canada to prioritize new HS Codes. In early 2007, Canada became the first country in the world to create special designations to track organic products moving across its border.

In December 2007, the Canadian federal government committed over 1.2 million Canadian Dollars to the organic sector, including 711'500 Canadian Dollars to help the industry's major research arm at the Organic Agriculture Centre of Canada in Truro, Nova Scotia, better identify research needs and more effectively communicate research results back to the sector. An additional 565'900 Canadian Dollars will be used to develop a national sector organization, the Organic Federation of Canada (OFC), to bring together all players in the industry, raise awareness of the sector and to help with regulatory development. Stephanie Wells, for OTA in Canada, holds the trade seat on the OFC and was named the first president of the federation. The remaining seats are for provincial and territorial organic sector representatives.

In addition, Agriculture and Agri-Food Canada has established the Organic Value Chain Roundtable (OVCRT), an industry representative body advising the government on sector-wide issues, including international branding and marketing, building domestic capacity, and regulatory issues. OVCRT is the only Roundtable that is not commodity-oriented. OTA in Canada is currently the OVCRT project-lead on a study of regulatory, trade and marketing barriers for Canadian organic products.

In early 2007, The Canadian Organic Growers launched the Growing Up Organic (GUO) project. The purpose of the GUO project is to build upon existing efforts to shift Canada towards increased organic production by exploring ways to increase the amount of organically grown food served in Canadian institutions, beginning with childcare centers. The primary motivation for starting with childcare centers relates to the fact that compared to other types of institutions, the demand for food is relatively small. Since lack of organic supply, particularly in some communities, is a key issue in Canada, COG wants to use childcare centers in order to gradually build the supply and distribution infrastructure required to serve larger institutions. Secondly, COG wants to ensure that the most vulnerable members of our society, young children, have access to organic food.

OTA in Canada

The Organic Trade Association (OTA), which has been active in Canada, stepped up its efforts in 2007 by hiring Matthew Holmes as the managing director of OTA in Canada. As a result, Holmes opened an OTA office in New Brunswick and divides his time between the eastern and Ottawa offices. In addition, Stephanie Wells, who has been key in initially developing the OTA in Canada effort, continues to represent membership and has opened a new OTA office in British Columbia. OTA maintains a close relationship with Canadian federal government departments, including the Canadian Food Inspection Agency and Canada Organic Office, Agriculture and Agri-Food Canada, the Department of Foreign Affairs and International Trade, Environment Canada, Industry Canada, and Health Canada.

References

Macey, Anne (2007): Retail Sales of Certified Organic Food Products in Canada 2006. Organic Agriculture Centre of Canada (OACC), Truro. Available at http://www.oacc.info/Docs/RetailSalesOrganic_Canada2006.pdf

Statistics Canada's 2006 Census of Agriculture, released in May 2007.

Certified Organic Production in Canada 2005, prepared by Anne Macey for the Canadian Organic Growers, August 2006. See www. www.cog.ca and http://www.cog.ca/OrganicStatistics.htm

North America: Organic Farming Statistics

Table 32: Organic agricultural land and farms in North America

Country	Organic agricultural area (ha)	Share of total agricultural land	Farms
Canada	604'404	0.90%	3'571
USA (2005)	1'620'351	0.50%	8'493
Total	2'224'755	0.57%	12'064

Table 33: Land use in North America

Land use details as of 2005

Main land use type	Main crop category	Agricultural area (ha)
Arable land	Cereals	382'261
	Fallow land as part of crop rotation	120'269
	Flowers and ornamental plants	24
	Green fodder from arable land	254'814
	Medicinal & aromatic plants	2'987
	Oilseeds	8'152
	Protein crops	96'328
	Root crops	3'160
	Textile fibers	49'095
	Vegetables	41'247
Arable land		*958'338*
Permanent crops	Citrus fruit	4'107
	Fruit and nuts	26'980
	Grapes	9'278
	Other permanent crops	4'956
Permanent crops		*45'321*
Permanent grassland		*991'024*
No information	*No information*	*230'071*
Total		2'224'755

Information Sources:

- Canada: Data provided by Anne Macey, Canadian Organic Growers (COG, www.cog.ca); Source: Information of the certifiers. Not for all certifiers data were available at the time of the survey
- US: Source: USDA Economic Research Service, Data Sets Organic Production at http://www.ers.usda.gov/Data/Organic/, Download of detailed data at http://www.ers.usda.gov/Data/Organic/Data/PastrCropbyState.xls

OCEANIA/AUSTRALIA

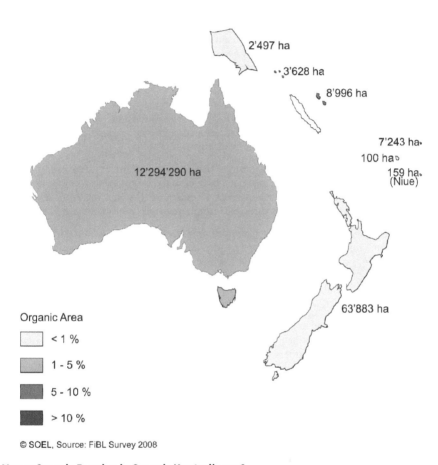

2'497 ha

3'628 ha

8'996 ha

7'243 ha

100 ha

159 ha
(Niue)

12'294'290 ha

63'883 ha

Organic Area

☐ < 1 %

▨ 1 - 5 %

■ 5 - 10 %

■ > 10 %

© SOEL, Source: FiBL Survey 2008

Map 7: Organic Farming in Oceania/Australia 2006

Organic Farming in Australia

ELS WYNEN[1]

Background

In the early 1980s, organic agriculture was of interest to two main groups in Australia. The first consisted of farmers, the second of regional and state-based organic gardening-farming organizations. Many of the farmers were geographically isolated and did not know of the existence of other organic farmers. The main reasons given by broadacre farmers for converting to organic agriculture was having experienced significant problems with their own or family's health or that of their crops or livestock when farming conventionally and feeling that drastic changes were needed to solve those problems (Wynen 1990).

The gardening-farming organizations usually operated in the capital cities of the six states, also in isolation, due to the large distances between cities in Australia. Although biodynamic farming was organized early on, in the 1980s a perceived need for cooperation and for combining the efforts of all forces in organic agriculture was growing.

In 1984, the idea of an umbrella organization that combined all forces interested in organic agriculture, including producers, consumers, traders, and researchers, was put forward. The National Association for Sustainable Agriculture, Australia (NASAA) was formally inaugurated in 1986, and incorporated in early 1987 (for more details, see Wynen and Fritz 2007).

Despite the overarching role NASAA gave itself, from the early 1990s it has concentrated on certification, which was also taken up by other organizations. The Biological Farmers of Australia (BFA) was started in 1987, and the Biodynamic Research Institute (BDRI) started to certify for the export market when the National Standards were introduced in 1992. Since this time, a number of other certifiers have started up (see next section).

In the late 1990s, the Organic Federation of Australia (OFA) came into existence. It was set up to unite all stakeholders in organic agriculture, as NASAA was in the early 1980s. At present, the OFA consists of a Main Board and several Member Councils, representing producers, consumers, certifiers, processors, traders, and the research and educational sectors. It makes policy decisions, lobbies government and other bodies on behalf of the organic sector, and represents the organic sector where appropriate. In 2007, it spent a considerable amount of its time and efforts on the advancement of organic standards for the domestic market (see below).

Size of the industry

Land under organic management

In the early 1990s, the area under organic management was estimated to be 150'000 ha for

---

[1] Dr. Els Wynen, Eco Landuse Systems, Canberra, Australia www.elspl.com.au

1990 (Hassall and Associates 1995). The estimate for 2006 is 12.3 million ha (Ian Lyall, AQIS, personal communication, November 2007), representing 2.8 percent of the total agricultural area of 440 million hectares in Australia (2003/2004), for which 1'710 producers were certified. This is a slight increase of area under organic management (11.7 million ha in 2005 and 12.1 million ha in 2004), and a decrease of number of farmers (1'894 in 2005).

Product range

The dramatic increase in area in the last decade is mainly due to certification of pastoral (extensive beef grazing) areas. Other important areas of production include: grains (wheat, rye, barley, oats, rice and oil seeds); fruit and vegetables, which are produced all year around; wine; dairy products; sheep, both for meat and wool; and herbs.

Land use

Figures from the two largest certifiers in Australia indicate that approximately 97 percent of the total certified area was under extensive grazing management in 2005.[1] This means that, of the total of 11.8 million hectares in that year, close to 370'000 hectares was in non-pastoral areas, which was approximately 0.7 percent of the total conventional area for those industries.[2] Although the non-pastoral certified organic area was only 3 percent of the total certified area, more than half of the total value of the organic sector originates from those areas.

Farm gate value

Wynen (2003) estimated that, in 2000-2001, only 38 percent of the total farm income of 89 million Australian Dollars (AUD)[3] (including organically grown products sold on the conventional market) was received for beef and sheep products, with around one quarter each for grains and horticulture. That is, the broadacre (grains, oilseeds) and horticultural sector accounted for more than half of the total value of the organic production in that year. A few years later, Halpin (2004) estimated total farm gate value of organic produce in 2003 (sold in the organic and conventional market) to be 140 million AUD.[4] Of the value for the products sold on the organic market (127.9 million AUD[5]), 40 percent accounted for beef, close to the estimate of the previous study. Also in this study, fruit, vegetables and grain made up about half of the total organic sales.

In summary, even though a large part of the area under organic production in Australia is used for extensive livestock production, products grown on less extensively-farmed areas have always been very important in organic production in Australia, accounting for at least half of the total value of the organic sector.

---

[1] Thanks are due to NASAA and the BFA for providing data.
[2] The total for wheat and other crops, mixed broadacre, and dairy for 2003/04 was 60 million hectare. It does not include the horticultural sector.
[3] 1 Australian Dollar (AUD) = 0.83858 US Dollars = 0.61227 Euros. Average exchange rate for 2007.
[4] This figure was a rough average of three years, estimated by the producers in a survey including 26 percent of all certified organic farmers.
[5] This figure was an estimate by adding all enterprises reported by the producer respondents.

Certification, Standards, Legislation

Accreditation of private certifiers

Europe has always been a major market for Australian organic produce. The introduction of Council Regulation (EEC) 2092/91 in 1991 altered requirements for imports of organic products, which meant that official certificates were required to accompany imports into the EU. To meet these requirements, government accreditation of organic certification organizations became necessary, and the Australian government (through the Australian Quarantine and Inspection Service AQIS) became involved in the accreditation of the private certifiers.

Since the 1990s, other certification organizations than NASAA, the Biological Farmers of Australia BFA (the certification arm of which is now called Australian Certified Organic (ACO), wholly-owned by the BFA, and a recently acquired non-AQIS accredited organization, the Organic Growers Association (OGA)) and the BDRI emerged. These include the Tasmanian Organic-Dynamic Producers (TOP), the Organic Food Chain (OFC), Safe Food Queensland (SFQ) and AusQual (2007).

Of the seven currently AQIS-approved certifying organizations, four are listed under European (and Swiss) law, and as such can provide inspection and certification services for all Australian export consignments to the EU. The same four organizations provide inspection and certification services for products exported to Japan, and three organizations have 'conformity assessment' arrangements with the National Organic Program (NOP) of the United States Department of Agriculture USDA (see table), with AusQual having applied for NOP recognition (November 2007).

Other countries, such as New Zealand, Malaysia, Thailand, Singapore and Canada, currently accept Australian 'certified' produce that has been issued a government organic export certificate to verify its authenticity. The Korean Food and Drug Administration (KFDA) recognizes ACO, BDRI, NASAA and OFC for processed organic foods. No AQIS-approved certifier is recognized in its own right for fresh organic produce. At present, no foreign certification bodies are operating in Australia, and no local certification bodies work in association with international certification bodies for certification within Australia (Jenny Barnes, AQIS, personal communication, November 2007).

Australian organic certification bodies and their legal export possibilities

		EU/Swiss	US	Japan
ACO	Australian Certified Organic	Yes	Y	Y
AusQual		No	Yes?	No
BDRI	Biodynamic Research Institute	Yes	No	Yes
NASAA	National Association for Sustainable Agriculture, Australia	Yes	Yes	Yes
OFC	Organic Food Chain	Yes	Yes	Yes
SFQ	Safe Food Queensland	No	No	No
TOP	Tasmanian Organic-Dynamic Producers	No	No	No

Australian certification bodies have their own standards, which are at least as stringent as the National Standard. For export purposes, AQIS is the accrediting body, that is, AQIS has the task to ensure that the certification bodies certify according to standards at least at the level of the National Standard, and according to accepted rules of complying.

National Standard for Organic and Biodynamic Produce

Organic production and processing in Australia has been prescribed by the National Standard for Organic and Biodynamic Produce since 1992; this National Standard was amended in 1998, 2002 and revised again in 2005. It stipulates the requirements for crop and landless plant production, animal husbandry, aquaculture, food processing, packaging, storage, transport and labeling, as well as complementing Australia regulatory requirements, such as environmental management and animal welfare (Organic Produce Export Committee 2002).

The National Standard is used for the purpose of export, and does not legally define 'organic' for the domestic market. This has been a source of two potential problems for the organic industry in Australia.

Although laws existed under the State/Territory fair trading acts (which draw their legal standing from the National Trade Practices Act) under which those who sell non-certified organic produce could be legally challenged on the basis of false and misleading labeling, success under this process was not likely (see Wynen and Fritz 2007). No other law protected the consumer of organic produce against false labeling. Although the export standards served as the de facto domestic standards, non-organic products could be sold as organic in the domestic market, as could products certified to standards not fully complying with the National Standard,[1] or by organizations not accredited by AQIS.[2] The second problem was that, due to WTO rules relating to national treatment, the Australian government could not prohibit imports of products labeled as organic, even if not produced according to the Australian Standard – or any other standard for that matter.

Australian Standard for Organic and Biodynamic Produce on the way

Since the early 1990, the organic sector has tried to achieve legal protection for the word 'organic,' but little progress was made until 2007. Early in 2007, the Organic Federation of Australia (OFA) applied with Standards Australia (an independent, not for profit body recognized by the Australian government as the standard-setting body) to develop an Australian Standard for Organic and Biodynamic Produce. After extensive consultation with key stakeholders, Standards Australia decided to take on this project, and instituted a Committee with the main stakeholders represented. It is expected that the Australian Standard, based on the National Standard employed since the early 1990s for the export market, will be adopted late 2008, and that a compliance scheme will accompany this Standard.

Once the standards and compliance scheme are in place, it will facilitate the prosecution of fraud and misrepresentation on the domestic market as well as the refusal of import of products deemed not to be equivalent to Australian domestic requirements. The same standards can also be used for the export market.

---

[1] ACO (Australian Certified Organic) certifies to 'Domestic-Only' (ACO 2005, 2006, BFA 2005, 2007).
[2] From 1 July 2007 OGA (Organic Growers of Australia) certifies small farmers (OGA 2007). It is not accredited by AQIS.

Market

Domestic Market

In the late 1990s, organic products were reported to account for only 0.2 percent of food retail sales nationally (Invest Australia and KPMG 1999, p.15). Only a few consumer studies are undertaken in Australia. Results of some show that, while there appears to be some positive correlation between income and the demand for organic food, no clear delineations can be made with respect to the consumption of organic food according to gender, income, age or education (Queensland Department of Primary Industries QDPI 2002; Smith 2003). Lockie and Donaghy (2004) found, however, that consumers of organic produce were more likely to be women, educated, and have at least middle-level incomes. They also reported that '...the attitude that stands out to many consumers in relation to organic systems is the perceived opportunity they offer for improved environmental outcomes, but that the premiums were higher than many were willing to pay. Authors of earlier studies cite price as an obstacle to a more rapid expansion of the Australian market for organics, in addition to quality concerns, availability, inconsistent labeling, and product recognition (Dumaresq & Greene 1997; Invest Australia and KPMG 1999; Lyons et al. 2001).

Current market figures for Australian organic produce are not available, and industry figures therefore need to be treated with caution. Farm-gate values for organic products in the early 2000s were estimated to be around 100 million AUD[1]. Wynen (2003) estimated farm-gate values in 2000-2001, including organic produce sold as conventional, at 89 million AUD, and Halpin (2004, p.20-21) - excluding organically grown produce sold as conventional - at 127 million AUD for 2003. However, estimates of retail values differ greatly, varying from less than 100 million AUD for 2000- 2001 (Wynen 2003) to 250 million AUD (BFA 2003), and 400 million AUD at which NASAA put the retail value in 2003.

The only commodity in which some more research has been undertaken recently is beef (Wynen 2006). This market has grown considerably since the late 1990s, when the large retailers entered the market. Whereas in 2000/01 the value of the Australian certified organic beef was only 32 million AUD (farm-gate prices), with less than two thirds going to the organic market, by 2005 the estimated production had doubled to around 60 million AUD (farm-gate prices), with virtually all of the produce being sold in the organic market. About three quarters was estimated to be sold in the domestic market. Dominant export markets moved from Japan and the UK in the early 2000s to the US in more recent years.

Prices on the domestic market

On the domestic market, organic produce receives a substantial price premium over that of conventionally grown produce. For cereals and livestock products, price premiums were reported by AQIS (see FAO 2002) as ranging between 50 and 75 percent, while for fruit and vegetables the premium was said to be usually between 50 and 60 percent; although, price premiums of up to 100 percent were considered not to be uncommon (Bulletin 2001). Halpin and Brueckner (2004, p.70) report higher premiums in 2003. The weighted average price premium of all goods were calculated as being 80 percent, with several products scor-

---

[1] 1 Australian Dollar (AUD) = 0.83858 US Dollars = 0.61227 Euros. Average exchange rate for 2007.

ing over 100 percent, such as wholemeal flour, muesli, olive oil, spaghetti (the highest at 287 percent), several vegetables (beans, zucchini, carrots), hard cheese and minced beef.

The pricing of organic food will continue to be a key determinant of consumer demand for organic produce and market growth, especially since it appears that current price premiums are set above levels many consumers accept (see for instance Pearson 2001; Queensland Department of Primary Industries QDPI 2003).

Exports

Exports of Australian organic produce have been mentioned as being 50 million AUD[1] (Austrade 2003). Europe is the key export market for Australian organic products, at least in quantities exported. Australia records its exports only in weights, not value. In 2001, Europe accounted for over 70 percent of Australian organic exports, with the main destinations being the UK, Italy, Switzerland, France, the Netherlands and Germany (Austrade 2003). More recently, although Europe is still the main market in quantity exported, the significance of the individual countries has changed somewhat. Especially France and Belgium are becoming more important, but other countries such as Japan, the US, Singapore, and Hong Kong have emerged as promising future export markets for Australian produce (Halpin and Sahota 2004, p.110). The primary products for export in 2003 were, in decreasing order of importance of quantity: grains; processed products; drinks and juices; and meat products. However, in terms of value, the order may well be different, and the importance of export destinations for Australia may also be different if values only are considered.

Imports

Australia also imports organic products, though the total value of imported organic produce is unknown. According to McCoy and Parlevliet (2000, p.62) imports in the late 1990s were mostly of processed grocery lines, such as coffee, pasta sauces, olive oil, soy drinks, preserves and the like, primarily from the UK and the US. Crothers (2003) reported in 2003 that some commodities were imported to fill temporary shortfalls in domestic production, such as kiwi fruit and fresh produce from New Zealand. For 2003, Halpin and Sahota (2004, p.112) estimated imports valued at 13 million AUD, with the main sources being New Zealand, the US and the UK. Products include not only food and drinks, of which more than half is processed, but increasingly non-edible items such as cotton and personal care products are imported.

Policy Support

As Australia's agriculture in general is export oriented, growth in the organic sector has been strongly influenced by rapidly growing overseas demand in the past. In recent years, the domestic market has probably increased in relative importance, due to increasing domestic demand, and severe drought conditions in extensive parts of agricultural areas severely curtailing production.

There is little government support to encourage organic agriculture per se. However, over

---

[1] 1 Australian Dollar (AUD) = 0.83858 US Dollars = 0.61227 Euros. Average exchange rate for 2007.

the recent past, governments both at Commonwealth and state levels have been supportive of the Australian Standards issue, and it seems likely that supporting regulations will be passed to make the standard effective on the domestic market. In addition, the Australian Competition and Consumer Commission (ACCC) has made funding available to assist in the development of the Australian Standard and promote an understanding amongst consumers.

Accreditation services are provided (through AQIS), although the certification organizations pay 60 percent of the cost of these services – 105'000 AUD for 2006/07 (Ian Lyall, AQIS, personal communication, November 2006).

The Organic Federation of Australia has also received government assistance with its Business and Marketing Plan, launched in 2007.

Many possibilities exist for government assistance in the farming sector in general, to help with developing innovations, overcoming marketing problems or attending courses, etc. These are detailed in DAFF (2004, Chapter 9), but most are available to all, not specifically organic farmers.

Research and extension

There is one research program (part of the Rural Industries Research and Development Corporation) devoted to organic agriculture since 1996 that has made available up to 270'000 AUD[1] per year to research and extension. For the next five years, this amount can be increased to a maximum of 450'000 AUD if the most favorable circumstances occur, where co-funding from other institutions happens – which seems not likely at present.

Most of the six state departments of agriculture have at least one officer dedicated to organic agriculture. Three states (Tasmania, New South Wales and Queensland) now have Ministerial Advisory Committees.

A combination of private funding has resulted in research on the size of the market, to be published early 2008.

Milestones

The organic sector seems finally poised to make progress in its endeavor to establish organic standards that can be used both for domestic and international trade purposes, through the setting of organic standards via Standards Australia (see section on 'Certification'). Standards Australia, an independent not for profit body, has set up a committee consisting of the main organic stakeholders involved in developing the Australian Standard – expected to be ready late 2008.

References

Austrade (2003), 'Organics Overview - Trade Statistics.' Retrieved 4th November, 2003, from

---

[1] 1 Australian Dollar (AUD) = 0.83858 US Dollars = 0.61227 Euros. Average exchange rate for 2007.

http://www.austrade.gov.au/australia/layout/0,,0_S2-1_CLNTXID0019-2_-3_PWB1106308-4_tradestat-5_-6_-7_,00.html.

BFA (2003), The Organic Food and Farming Report 2003, Biological Farmers of Australia, Toowoomba.

Bulletin (2001), Organic Growth Industry, The Bulletin, 18 April.

Crothers, L. (2003), Australia - Exporter Guide Annual 2003, USDA Foreign Agricultural Service, Canberra.

DAFF (2004), 'Government resources and assistance' in DAFF, The Australian Organic Industry: A Profile, Department of Agriculture and Fisheries, Canberra.

Dumaresq, D. & Greene, R. (1997), 'Overview of the organic industry in Australia' in D. Dumaresq, R. Greene & L.van Kerkhoff (Eds.), Organic Agriculture in Australia, No. 97/14, pp. 69-80, RIRDC, Canberra.

FAO (2002), 'Organic agriculture, environment and food security. Environment and Natural Resources Series No. 4, Rome.

Halpin, D. (2004), 'A farmlevel view of the Australian organic industry' in DAFF, The Australian Organic Industry: A Profile, Department of Agriculture, Fisheries and Forestry, Canberra.

Halpin, D. and Brueckner, M. (2004), 'The retail pricing, labelling and promotion of organic food in Australia' in DAFF, The Australian Organic Industry: A Profile, Department of Agriculture and Fisheries, Canberra.

Halpin, D. and Sahota, A. (2004), 'Australian food exports and imports' in DAFF, The Australian Organic Industry: A Profile, Department of Agriculture and Fisheries, Canberra.

Hassall & Associates (1995) The domestic market for Australian organic produce in Australia: an update. Report prepared for RIRDC, Canberra.

Invest Australia and KPMG (1999), 'Value Added Organic Food Study: Benchmarking the Competitive Strengths of Australia for Organic Products'. Unpublished report.

Lockie, S. and Donaghy, P. (2004), 'Who consumes organic food in Australia' in DAFF (2004), The Australian Organic Industry: A Profile, Department of Agriculture and Fisheries, Canberra.

Lyons, K., Lockie, S. & Lawrence, G. (2001), Consuming 'green': the symbolic construction of organic foods, Rural Sociology 11(3), 197-210.

McCoy, S. & Parlevliet, G. (2000) 'Export market potential for clean and organic agricultural products.' Report No. 00/76, RIRDC, Canberra.

NASAA (2003), Australian research, Technical Bulletin for NASAA Certified Operators 1(4), 11.

Organic Federation of Australia (2007). Corporate Plan. North Sydney. Available at http://www.ofa.org.au/papers/OFA_Corporate_Plan_2007.pdf

Organic Produce Export Committee (2002), National Standard for Organic and Bio-dynamic Produce, AQIS, Canberra.

Pearson, D. (2001), How to increase organic food sales: results from research based on market segmentation and product attributes, Agribusiness Review 9(8).

Queensland Department of Primary Industries (QDPI) (2002), Queensland organic buyer profile. Retrieved 3rd October 2003, from http://www.dpi.qld.gov.au/attitude_bites/12246.html.

Queensland Department of Primary Industries (QDPI) (2003) 'Drivers of Consumer Behaviour. Organic Food'. Retrieved 2nd October 2003, from www.dpi.qld.gov.ai/businessservices/11480.html.

Smith, J. (2003), Consumer sentiment ripe for new food category, Food Week 1-2, 7 July.

Wynen, E. (1990), 'Sustainable and Conventional Agriculture in South- Eastern Australia - A Comparison'. Economics Research Report No.90.1, School of Economics and Commerce, La Trobe University, Bundoora. See http://www.elspl.com.au/abstracts/orgprodgen2001.HTM for a summary.

Wynen, E. (2002), 'What are the key issues faced by organic producers?' In Organic Agriculture: Sustainability, Markets and Policies. Proceedings of an OECD Workshop held in Washington DC, United States, 23-26 September 2002, Paris. http://www.elspl.com.au/abstracts/abstract-a18.HTM.

Wynen, E. (2003), Organic Agriculture in Australia - Levies and Expenditures. Report No. 03/002. RIRDC, Canberra (http://www.rirdc.gov.au/reports/org/02-45.pdf).

Wynen, E. (2006), 'Organic beef production and marketing in Australia', Journal of Organic Systems, 1(1). http://www.organic-systems.org/journal/0601/pdf/wynen.pdf.Wynen, E. and Fritz, S (2007), 'NASAA and organic agriculture in Australia', in W. Lockeretz (ed.), Organic farming – an international history, CABI, Wallingford, UK, pp. 225-241. Retrieved 10 December, 2007, from http://www.elspl.com.au/OrgAg/Pubs/Pub-A-FP/OA-FP-A24-Lockeretz-Chapter13.pdf.

215

Organic Farming in New Zealand

SEAGER MASON[1]

Introduction

Organic agriculture and food production in New Zealand has developed steadily over the last 25 years. The most rapid growth has been since the mid 1990s, driven by various factors such as market demand for organic products, opposition to genetic engineering, and other environmental and food safety concerns. There is wide recognition in NZ of the important role that organic agriculture can and does play in moving agriculture and food production towards more sustainable practices, as well as recognition of the value of producing high quality certified organic products for export markets and the domestic market.

Through the establishment of the sector umbrella organization Organics Aotearoa New Zealand (OANZ) and the Organic Advisory Programme (OAP), as well as other initiatives, there is political recognition of the benefits of organics. This recognition is of the commercial potential of organics, the importance of organics in helping to underpin New Zealand's clean green image as a producer of high quality agricultural products and as a tourist destination, and also of the benefits for health and the environment.

Statistics 2007

A recent study of New Zealand's organic sector was commissioned by Organics Aotearoa New Zealand (OANZ), and carried out by the Centre for the Study of Agriculture, Food and Environment (CSAFE), University of Otago. The Summary Report of this study was released August 2007 (Grice et al. 2007). Most of the statistics quoted below are from this study.

The main types of organic primary products in New Zealand are apples, kiwifruit, blueberries, fresh and processed vegetables, arable crops, dairy products, meat and wool, viticulture and aquaculture products. The biggest organic sectors so far are apples and kiwi fruit, with ten and five percent of the total production respectively.

The current growth sectors for organic primary production are apples, livestock (particularly lamb and wool), and viticulture. There is strong interest in organic and biodynamic viticulture with a steady increase in the number of vineyards converting. There is rapid growth in the number of livestock farms converting in response to very good premiums for organic lamb for export markets and demand for organic wool.

Some current statistics

- Certified producers: 860 certified organic producers with 1206 certified organic operations in total.
- Certified land area: 63'883 ha certified organic land.

---

[1] Seager Mason, Technical Director of BioGro New Zealand Inc., Wellington, New Zealand, www.bio-gro.co.nz. BioGro NZ is New Zealand's largest and leading organic certifier and organic producers' organization.

- Exports: 120 million NZD,[1] growing at more than 10 % per annum.
- Domestic market: More than 200 million NZD, growing at more than 10 % per annum. 59 % of this is produced in New Zeeland, and 41 % is imported. This is 1.1 % of the New Zealand market for food and beverages.
- Kiwifruit: Organic production is approximately 5 % of the kiwifruit sector.
- Apples: Organic production is approximately 10 % of the apple sector.
- Vegetables/cropping: Organic production is about 2 % of the sector.
- Dairy and meat: Organic production is still less than 1 % of the sector.
- Certifiers (approximate numbers): BioGro: 480 producers; Demeter: 40 producers; Organic Farm New Zealand (small scale producers scheme): 150 producers; Agriquality: 190 producers.

Markets

Domestic

New Zealand's domestic market grew very rapidly over the period 2000 - 2002, by more than 100 percent per annum each year. This growth was due to a variety of factors, but in particular because of:

- Rejection of genetic engineering;
- The increasing range and high quality of organic products on the market;
- Increasing number of outlets, particularly supermarkets, stocking organic products; and
- Many people wanting to support organic agriculture as being the best way forward for New Zealand's agriculture and food production.

Most food and beverage products are now available as certified organic, most supermarkets now stock at least some organic products, and some supermarkets are specializing in organics due to customer demand. Organic shops are increasing in number and size, with some of the successful organic shops becoming small to medium size organic supermarkets, and there are now some chains of organic shops. The domestic market continues to grow steadily, and part of the growth includes products other than food and beverages such as organic health and body care products, garments, and household cleaners.

Exports

New Zealand's economy is reliant on exporting, and agricultural products are New Zealand's main exports. Exports of organic products have grown steadily over the last 17 years, and are currently 120 million NZD per annum.

Exports by category

- Fresh fruit and vegetables: 73%
- Meat and wool:8%
- Dairy products: 6%

---

[1] 1 New Zealand Dollar (NZD) = 0.57165 Euros = 0.73608 US Dollars. Average Exchange rate 2007.

- Processed food: 5%
- Honey: 3%
- Wines, beers, juices:3%
- Other (aquaculture etc):2%

Exports by markets

- Europe: 46%
- US and Canada: 27%
- Japan: 12%
- Korea: 5%
- Other Asia: 5%
- Australia: 4%
- Other: 1%

Demand for exports of organic products in most sectors exceeds supply. A key focus of the Organic Advisory Programme is to provide primary producers with information and support for conversion to organics to increase New Zealand's organic production in order to meet this growing demand.

Standards and legislation

The New Zealand Standard for Organic Production

The New Zealand Standard for Organic Production was released in November 2003. It was developed with Government funding under the auspices of Standards New Zealand. At this stage it serves as a benchmark for certifiers operating in the domestic market. It is a voluntary standard, not mandatory, so consumer protection is through the Fair Trading Act, with reference to the New Zealand Standard as required. There are no specific organic labeling laws in New Zealand.

Export

Exports to the European Union and US are via the New Zealand Food Safety Authority (NZFSA) Official Organic Assurance Programme (OOAP). Through this program, New Zealand has Third Country Listing with the European Union, and recognition for the USDA National Organic Program of the US Department of Agriculture (USDA). The export certifiers such as BioGro operate as Third Party Agency certifiers for the OOAP.

Exports to Japan have two options, either through the export certifier such as BioGro having RFCO (Recognized Foreign Certification Organization) status with the Japanese Ministry of Agriculture MAFF, or through NZFSA OOAP equivalence with JAS Organic.

Exports to Quebec are through the export certifier having recognition with CAAQ (Conseil des Appellations Agroalimentaires du Quebec).

Exports to other markets are through meeting the requirements of that market, such as certification by an IFOAM accredited certifier.

Imports

There are no controls on imports labeled 'organic' other than certifiers setting their own standards for recertification, and through the Fair Trading Act.

State Support

There is a small amount of Government support for organics in New Zealand. The main recent examples are:

- New Zealand Standard for Organic Production: see above
- New Zealand Organic Sector Strategy: A Government funded Organic Sector Strategy was released in November 2003. A key recommendation was for the formation of a peak industry body, Organics Aotearoa New Zealand (OANZ), to coordinate initiatives in the organic sector. The strategy has set an ambitious target of on billion NZD[1] worth of sales by 2013.
- Organic Farm New Zealand: This is a scheme for certification of small scale producers, which was developed by the Soil & Health Association with Government funding. The scheme is based on 'pods' (groups) of producers, regionally based, with each pod able to operate their own certification system, but linked to a national coordinating body. Through voluntary input, this provides low cost certification for small scale producers. This scheme continues to be administered by the Soil & Health Association.
- Organics Aotearoa New Zealand (OANZ): OANZ was launched in November 2005 with Government funding for at least three years to establish.
- OAP: The Organic Advisory Programme also has Government funding, and is operated by OANZ. The OAP provides information and support to producers and processors considering conversion to organics, and also to those already in conversion or with existing organic operations.

Research and Extension

Organic research in New Zealand is carried out mainly by crown research institutes, universities, and the private sector. One example is an organic research farm that is a joint venture between a university and a food processing company. There are also some producer groups such as in the organic kiwifruit, pipfruit, dairy, viticulture, and avocado sectors, which have significant input into coordinating research and extension. In general, the view is that research funding for organics is inadequate, particularly as developments in organics typically benefit conventional production also. It is well recognized that much of the knowledge base in organics is with the experienced producers, and some of the 'research' happens on farm as successful farmers develop their production systems. Organics Aotearoa New Zealand (OANZ) is now taking a coordinating role for organic research.

Several universities and other tertiary institutions, as well as some private organizations, offer courses and training in organics. There are an increasing number of agricultural advis-

---

[1] 1 New Zealand Dollar (NZD) = 0.57165 Euros = 0.73608 US Dollars. Average Exchange rate 2007.

ers who offer consultancy services for organic producers, mainly through the Organic Advisory Programme.

Outlook

Political

Through the launch of the New Zealand Organic Sector Strategy and the establishment of Organics Aotearoa New Zealand (OANZ), there is Government acknowledgement of the importance of organics in New Zealand, but still there is only very limited Government support compared to other sectors.

Organic organizations such as the Soil & Health Association take a very active public role on issues such as food safety and the environment.

Genetic Engineering

Genetic engineering (GE) continues to be a major issue in New Zealand, and was the number 1 issue in the general election in July 2002. There was a moratorium on commercial release of genetically modified organisms until October 2003, but in spite of majority public and industry support for it to remain, that was lifted. There is currently one field trial of GE herbicide resistant onions in New Zealand, and an application to field trial GE Bt brassicas was approved, but this decision has been appealed by GE Free New Zealand with support from OANZ, BioGro, and the Biodynamic Farming and Gardening Association. Such applications are strongly opposed by the organic sector and many other organizations and people. There is a very active movement for New Zealand to remain non-GE, and this is supported by a majority of New Zealanders. GE remains an important issue for New Zealand's organic sector.

Other

A key issue for New Zealand's organic sector is lack of production to meet growing demand, both for the export market and the domestic market. The only solution is to encourage more farmers and growers to convert by providing advice and research to support conversion. Organics Aotearoa New Zealand (OANZ), the various sector and regional organic organizations, and the established organizations such as BioGro, Soil & Health Association, and the Biodynamic Farming and Gardening Association are working hard to facilitate this support.

Reference

Grice, J., Cooper, M., Campbell, H. and Manhire, J. The State of the Organic Sector in New Zealand, 2007- Summary Report. Centre for the Study of Agriculture, Food and Environment: University of Otago. (2007). Available at http://www.csafe.org.nz/Organics%20Summary%20Report.pdf

Organic Agriculture in the Pacific[1]

KAREN MAPUSUA[2]

The Pacific region comprises of a number of island countries, and includes the Cook Islands, the Federated State of Micronesia, Fiji, Kiribati, the Marshall Islands, Nauru, Niue, Palau, Papua New Guinea, Samoa, the Solomon Islands, Tonga and Tuvalu and Vanuatu. The Pacific islands region has a collective population of approximately 8 million people, and a combined island land mass of 525'000 square kilometers in a sea area of more that 14'000'000 square kilometers.

Pacific Island countries, while highly diverse, share common constraints that impede their efforts to achieve balanced economic growth and sustainable food security. Major constraints include small size, remoteness, geographic dispersion and vulnerability to natural hazards, as well vulnerability to external economic conditions. Most people, more than 80 % in some countries, live in rural areas and rely heavily on agriculture, forestry and fisheries. The export sector generally comprises a narrow range of primarily agricultural commodities.

Organic agriculture development has considerable potential in the Pacific region due to increasing demand for high-quality products, desire to protect the environment and biodiversity, and the farming family, but there are also limiting factors to consider. The overall quantity of organic production and trading is still very small at present. Furthermore, there is no legal framework on organic agriculture and no overall development strategy under which the main actors could cooperate with each other. The development of organic agriculture in the region might be characterized as driven by farmers' organizations and other NGOs, and subsequently taken up by government agencies. Gradually partnerships are being formed that might provide participatory development of the sector.

Organic agriculture development in the Pacific

Organic agriculture is not a new concept in the Pacific; it is very much the traditional farming system that Pacific forefathers practiced sustainably for centuries. Today, current farming practices in many communities are still based on 'age-old' systems that are free from the residues of agrichemicals and where environmental integrity remains largely intact. However, the motives for organic farming have changed. In the past, farming was predominantly for subsistence living, but in the cash driven societies of today, there is now a need from overseas markets to ensure that products being labeled and sold as organic produce meet international standards.

Organic agriculture provides important opportunities for Pacific Island Countries (PICs) to export to niche markets a number of high-value, low-volume crops and enhance economic sustainability. Meeting the requirements of international certification has posed a number

---

[1] This report is an abridged version of the Desk Survey of Organics in the Pacific Region published by IFOAM, 2007
[2] Karen Mapusua, Associate Director, Women in Business Development Inc., PO BOX 6591, Apia Samoa, www.womeninbusiness.ws

of problems for the countries, which include, but are not limited to:

- the relative high costs associated with attaining and maintaining organic certification;
- lack of government awareness of the potential benefits of organic agriculture as opposed to currently practiced alternatives; and
- the lack of broad based and supporting policies on organic agriculture.

The rationale for the development organic agriculture should also include promotion of local consumer awareness about the benefits derived from the consumption of organic products and assist with the development of domestic markets including tourism markets.

Apart from the market opportunities, organic agriculture is relevant for the Pacific with regard to the promotion of self-reliance, and will effectively address food security and food sovereignty issues, and in particular, the possible reduction of food imports.

Regional and National Policy & Frameworks

Regional level

Presently, a regional policy does not exist specifically for organic agriculture; there are a number of other regional frameworks that mention sustainable livelihoods, but even within these frameworks there is limited reference to organic agriculture and its potential value to PICs. The extent to which organic agriculture issues and benefits are understood by policy-makers is still very much at the elementary stage.

However, the Pacific Islands Forum Secretariat (PIFS) formally endorsed the Regional Strategy for Agriculture Development and Food Security, and the Regional Program for Food Security (RPFS) in the Forum Island Countries (FICs). This document drew attention to weaknesses in policy and program formulation capacity with respect to addressing food security at both national and regional levels. The Regional Program for Food Security (RPFS) follows a three-pronged approach, which would help FICs to adjust to changes in the international trade environment brought about largely by the Uruguay Round Agreement on Agriculture (URA). The three areas where actions would be supported relate to:

a) trade facilitation, which would improve the environment for and remove impediments to trade;

b) policy harmonization, which would help create a conducive domestic policy framework for promoting efficient production systems in line with comparative advantage; and

c) community level investments, which would allow farmers to adjust to, and take advantage of, new opportunities resulting from trade facilitation and policy harmonization.

In terms of organic agriculture, the strategy notes the "need to diversify the export base away from traditional commodities has been acknowledged by all governments who are committed to facilitate the upgrading of the production and the exploitation of niche markets. Potential export commodities might include...certified organic produces..."

While this is a positive statement, the strategy has not fully realized the potential benefits of organic agriculture and the positive implications of providing resources towards its development.

National level

Nearly all Governments are generally sensitized to environmental issues and possess some information on economic and social benefits of organic agriculture, although many of them do not have any government policy on organic agriculture in place. Some Governments are demonstrating an increasing attention to natural agriculture, and in particular organic agriculture. In Samoa, a National Organic Advisory Committee has been created under the chairmanship of the Prime Minister and in collaboration with a local NGO to drive and promote the development of organic farming in the country. This committee represents an interesting practice of government collaboration with civil society that could be replicated in other countries to lead to shared and effective strategies for the promotion of organic agriculture. There is some movement towards the development of organic agriculture in several countries, as is demonstrated by the national initiatives shown in the following table.

Organic Policies and Standards

Government	Legislation and Policies on organic agriculture	National Standards	National Initiatives
Cook Islands	No	Cook Islands National Standards for Organic and Biodynamic Produce (2001). Endorsed by CIOA and Ministry of Agriculture	Training programs conducted on organic crop production methods
Fiji	No	No	Organic agriculture considered priority for the diversification of production and exports
Niue	Niue National Strategic Plan and Niue Environment Act 2004 refer to organic agriculture	No	Niue National Strategic Plan and Niue Environment Act refer to organic agriculture as means to promote the growth of agriculture; close collaboration with NIOFA
Kiribati	No (only a law prohibiting the use of chemical fertilizers)	No	Training activities; support to KOFA
Papua New Guinea	No	No	Research; training
Samoa	National Organic Advisory Committee (chairmanship of the Prime Minister) National Organic Strategic Plan 2005	No	National Organic Advisory Committee
Solomon Islands	No	No	Explore market opportunities for organic agriculture products

Government	Legislation and Policies on organic agriculture	National Standards	National Initiatives
			agriculture products
Tonga	National Organic Certification system (NOC) and Organic Coordination and Development Committee (OCD)	No	Development of local certification systems
Vanuatu	No	No	Subsidize livestock certification

Organic farmers' associations

Since the beginning of the year 2000, several national associations of organic producers and farmers were founded in the region with the aim to organize the individual organic producers in each country. Many of these associations have not developed into well functioning organizations for a number of reasons. Farmers were too busy or geographically dispersed, and they were also not informed enough to aid in the development and management of the organizations.

Non-governmental organizations (NGOs)

The NGOs supporting organic farmers establish a relationship with farmers' groups and provide them with assistance, playing a crucial role in implementing activities related to the organic process, from training, technical and financial support to certification and supporting the development of Internal Control Systems. Several local and international NGOs have promoted organic agriculture as an appropriate technology for small-scale farmers in the Region. NGOs emphasize some aspects of organic agriculture, such as the limited use of inputs, its independence from agri-business and its care for natural resources, as well as the potential for food security, economical viability and gender equity. Some NGOs, including Women In Business Development Inc. (WIBDI), view organic agriculture more from a market perspective.

Markets

Export markets

Most of the organically certified products from the Region are for export. Due to their proximity and the presence of large communities of Pacific Island emigrates, the main international markets are Australia and New Zealand. Japan, North American and Europe are other growing markets. The following summary table lists the main crops that are currently organically certified and exported from the Pacific region.

Domestic markets

Generally, the domestic markets for certified organic products are not very developed, and in some cases are non existent. Organic products are commonly sold as conventional without a premium price. Some initiatives are ongoing or are in the pipeline to promote the consumer awareness about organic products. Ecotourism in several countries, in particular for Fiji, Cooks and Samoa, holds a great deal of potential.

Porducts grown for export in the Pacific Islands

Products grown for export	Countries
Vanilla & other spices & nuts	Fiji, Vanuatu, Niue
Cocoa	Vanuatu, Samoa, PNG, Vanuatu
Virgin Coconut Oil	Samoa, Fiji, Solomon Islands
Nonu /noni (Morinda Citrifolia)	Cook Islands, Samoa, Fiji, Niue
Banana, Guava & Mango	Fiji, Samoa
Taro	Cook Islands
Papaya (pawpaw)	Cook Islands, Fiji
Bananas	Fiji, PNG
Coffee	PNG
Beef	Vanuatu

Outlook

The traditional farming systems of the Pacific region are poised for conversion to organic; many of the subsistence farmers have never applied commercial fertilizers. There is a widespread perception by farmers and governments that organics holds great potential. Organic agriculture is also being investigated by universities and other competent agencies in the region. Organic aquaculture, sustainable forestry; sustainable fisheries and sustainable tourism are generating interest by governments throughout the region, and there is full support from local stakeholders involved to collaborate in supporting regional development.

There are, however, constraints, such as the small size of the organic sector in the region and a lack of regular and reliable supplies. There is a lack of education in rural areas across the region, especially lack of knowledge about certification. There are also few resources and funds dedicated towards extension, education and promotion projects on organics. The differences in the development and knowledge of organic agriculture in the region may provide challenges in developing a regional approach; there is a need for greater organization, coordination and cooperation within the organic movement.

There is a need to increase the public awareness about organic agriculture, which could lead to development of domestic markets for organic products. While the quality of products for export is generally good, there is a lack of adequate facilities for packaging and processing. Another significant issue is that national and regional associations of organic producers are in the early stages of development, and capacity building efforts to improve organic farming practices, create an effective business environment and coordinate overall activities are necessary to bring the organic sector forward. While national governments appear interested in developing the organic sectors, most still lack policies and regulations.

References

SPC, 2006. Heads of Agriculture and Forestry Services Final Resolutions, power point presentation, Secretariat of the Pacific Community, Suva Fiji

PIFS, 2002. The Regional Strategy for Agriculture and Food Security in the Forum Island Countries. Pacific Islands Forum Secretariat, Suva Fiji

Oceania/Australia: Organic Farming Statistics

Table 34: Organic agricultural land and farms in Oceania/Australia 2006

Country	Organic agricultural area (ha)	Share of total agricultural land	Farms
Australia	12'294'290	2.80%	1'550
Fiji (2005)	100	0.02%	No data
New Zealand	63'883	0.37%	860
Niue	159	1.99%	61
Papua New Guinea	2'497	0.24%	4'558
Samoa	7'243	5.53%	213
Solomon Islands	3'628	3.10%	352
Vanuatu	8'996	6.12%	No data
Total	12'380'796	2.70%	7'594

Source: FiBL Survey 2008

Table 35: Organic agricultural land use in Oceania/Australia 2006

Main land use type	Agricultural area (ha)
Cropland, other/no details	368'99
Permanent grassland	11'925'461
No information	86'406
Total	12'380'796

Source: FiBL Survey 2008

Information sources / Data providers

- Australia: Data provided by Els Wynen, Eco Landuse Systems, Flynn, Australia, http://www.elspl.com.au/, based on information of the certifiers
- Fiji: Fiji Organic Association
- New Zealand: Grice, Janet, Mark Cooper, Hugh Campbell & Jon Manhire (2007): The State of the Organic Sector in New Zealand, 2007. Summary Report August 2007. Centre for the Study of Agriculture, Food and Environment University of Otago. http://www.csafe.org.nz/Organics%20Summary%20Report.pdf
- Niue: Data provided by Karen Mapusua, Women in Business Development Inc. (WIBDI), Apia, Samoa, www.womeninbusiness.ws, Source: Certifier data
- Papua New Guinea: Data provided by Karen Mapusua; Source: Certifier data
- Samoa: Data provided by Karen Mapusua; Source: Certifier data
- Solomon Islands: Data provided by Karen Mapusua; Source: Certifier data
- Vanuatu: Data provided by Karen Mapusua; Source: Certifier data

Achievements Made and Challenges Ahead

ANGELA CAUDLE DE FREITAS[1] AND GABRIELE HOLTMANN[2]

When the founders of IFOAM met on November 5, 1972 at the Palais des Congrès in Versailles, France, they looked forward and focused on a vision of a worldwide united organic movement. The chosen name 'International Federation of Organic Agriculture Movements' reflected the future aims. They saw the major need of the international organic sector: the diffusion and exchange of information on the principles and practices of organic agriculture.

Thirty-five years later, the rapid global development of organics has brought with it more challenges and opportunities. With the steady growth in the organic sector, many actors have hopes and aspirations for the future of organic agriculture; there are highly variable perspectives about the purpose and the future of organic agriculture. **A strong, united movement for global Organic Agriculture has never been more important**. This is a critical time for innovative solutions to global challenges of environmental protection, food security, social justice, and health.

Achievements made

Currently, global trade relations and rules, international and national policies, structural adjustments and trade concentration affect the self-image of the organic movement. In response to this, IFOAM concentrated on specific topics in the past year. Organic trade, food security, rural development, marketing of organic and regional products, the revision of the IFOAM Organic Guarantee System, the East Africa Organic Products Standard and the IFOAM recommendation on group certification were some of the areas of focus for IFOAM.

Organic Trade 2007

Trade in organic products is booming. The world's largest supermarkets and discounters are launching lines of organic products, and big organic supermarket chains have started to negotiate with the conventional industry, seeing this as a faster way to achieve growth. To discuss the challenges and opportunities resulting from these changes, IFOAM organized the trade symposium *In-Between Discount and Premium - Friend or Foe?* in February 2007, one day before BioFach. The importance of taking a holistic approach in organic trade was stressed during this event. Dr. Götz Rehn, Founder and Director of Alnatura, a leading organic brand and chain of supermarkets in Germany, emphasized this notion, noting 'The aim of a business should be to produce good products that serve mankind and protect the earth. The achievement of profit is the expression and measure of how efficient this is being done.' Thomas B. Harding, Jr., President of AgriSystems International and former IFOAM

---

[1] Angela Caudle de Freitas, Executive Director of the International Federation of Organic Agriculture Movements (IFOAM), 53113 Bonn, Germany, www.ifoam.org.
[2] Gabriele Holtmann, International Federation of Organic Agriculture Movements (IFOAM), 53113 Bonn, Germany, www.ifoam.org.

President went on to say that 'Building a sustainable, high quality and ethical organic business is not about the size or scale of the operation. Crucially, it is about creating systems that deliver organic products that hold fast to the Principles of Organic Agriculture and impart holistic values to consumers, irrespective of scale.'

Organic Guarantee System

For more than 30 years, IFOAM has been developing the IFOAM Basic Standards (IBS). The IBS started out in IFOAM's early years as a collection of general principles and guidelines for organic agriculture, and evolved over nearly two decades into detailed standards that could be directly applied to organic production and processing. In response to the growing number of private and regional certification standards and as a result of a mandate by the IFOAM General Assembly, the role of the IBS has been transformed, and they have come to represent 'standards for standards.'

Since the mandate from the IFOAM General Assembly in 2005, IFOAM has been revising the Organic Guarantee System (OGS) to provide more access to it for producers, standard setting and certification bodies and governments, while maintaining organic integrity. An integral part of the OGS revision is the revision of the IBS, which started in late 2005; this revision continues in 2008, and is subject to the approval of IFOAM's members. The revision aims to fully transition the document to 'standards for standards.' A further aim is to establish the revised document as a benchmark for facilitating the equivalence of various organic certification standards, including private standards and those of governments (such as the US NOP and the EU organic regulation). Organic certification standards that meet this benchmark will be included in the IFOAM Family of Standards and recognized by IFOAM as equivalent. Governments and private certification bodies will be invited to also recognize standards in this Family as equivalent to their own standard. The result is a means to streamline the trade of organic products worldwide. IFOAM has lead the International Task Force for Harmonization and Equivalence in Organic Agriculture (ITF), which is a joint initiative of IFOAM, the FAO and UNCTAD. The tools for equivalence and recognition being developed during IFOAM's Organic Guarantee System Revision are parallel to tools and recommendations of the ITF. The ITF provides the needed bridge between the private and government to identify and use common tools for streamlining trade.

East Africa Organic Products Standard

A uniform set of requirements for growing and marketing organic produce were established for East Africa and the Prime Minister of Tanzania introduced them at a conference in Dar es Salaam at the end of May 2007.

The East African Organic Products Standard (EAOPS), the second regional standard in the world, was developed by a public-private sector partnership in East Africa, supported by the UNEP-UNCTAD Capacity Building Task Force on Trade, Environment and Development (CBTF), a joint initiative of the United Nations Conference on Trade and Development (UNCTAD) and the United Nations Environment Programme (UNEP), and IFOAM.

The EAOPS and associated East African Organic Mark will ensure consumers that produce so labeled has been grown in accordance with a standardized method based on traditional methods supplemented by scientific knowledge, and based on ecosystem management rather than the use of artificial fertilizers and pesticides. As organic produce generally sells at premium prices in rapidly growing overseas markets, it is hoped that the standard will increase sales and profits for small farmers in the region.

Food Security

At the beginning of May 2007, the Food and Agriculture Organization of the United Nations (FAO) held an international conference on Organic Agriculture and Food Security at its headquarters in Rome, Italy. IFOAM was a conference partner and helped to present the results to the Committee on World Food Security. The overall objective of the Conference was to shed light on the contribution of Organic Agriculture to food security through the analysis of existing information in different agro-ecological areas of the world. The conference identified Organic Agriculture's potential to address the food security challenges, including conditions required for its success.

The outcome of the Conference was a thorough assessment of the state of knowledge on Organic Agriculture and food security, including recommendations on areas for further research and policy development. The Report of the Conference was submitted to the 33rd Session of the Committee on World Food Security, which concurred with the conference recommendation that organic agriculture be included as an option in FAO's field program on food security .

Marketing of organic and regional values

At the end of August 2007, IFOAM organized the 1st International Conference on the Marketing of Organic and Regional values. Participants discussed ideas, opportunities and strategies to protect organic product identity, traditional knowledge and biodiversity, and thus farmers and rural communities. Dr. Vandana Shiva and Helena Norberg-Hodge, both winners of the Right Livelihood Award, addressed the importance of bringing back value in local and regional economies that are increasingly getting lost in our globalized world.

One of the concerns addressed was the policies that encourage inequitable competition between producers in industrial countries and those in developing countries, severely constraining production in developing countries as a result. The developed countries directly affect production in developing countries by dumping their agricultural surpluses in developing countries, which creates unfair competition resulting from perverse subsidies. When sold on the world market at less than the cost of production, these surpluses depress local prices, thereby lowering production and peoples' direct access to food.

Group Certification

A grower group is typically a group of smaller producers that is regulated by a quality control system, ensuring the group's compliance to a certain standard. IFOAM has established guidelines for setting-up internal control systems and for inspecting and certifying operators under a group certification scheme.

The Certification, Accreditation and Compliance Committee (CAC) of the National Organic

Standards Board (NOSB) of the United States Department of Agriculture National Organic Program (USDA NOP) drafted a recommendation concerning group certification. The US being the largest market for organic products in the world, and group certification being a main model through which developing country organic producers are certified, it appeared essential for IFOAM to take a leading position in influencing the outcome of the NOSB recommendation drafting process. The recommendation will have far-reaching implications for a considerable number of smallholders in developing countries, as well as for processors and traders already struggling with un-harmonized national organic regulations and certification requirements. Since August 2007, IFOAM liaised with those who have been drafting the CAC recommendations and with the US National Organic Program (NOP) to educate on and advocate for the group certification concept based on IFOAM's criteria and guidelines. Despite these efforts, IFOAM was not satisfied with the final CAC recommendations. Fulfilling its role of leading the organic movement worldwide, IFOAM worked on a response to the recommendation, and asked their members to take action to influence the outcome of the NOSB vote. IFOAM will continue to work with the NOP to ensure that group certification is allowed under the USDA NOP. Through this process, IFOAM will also work to further improve the current IFOAM criteria and guidance to ensure that major areas of concern in group certification are addressed.

Challenges ahead

The organic movement must address the current and emerging challenges, beginning by ensuring organic integrity of products coming from developing countries. These issues are strongly linked to the exceptionally high growth rates that have created a high demand for organic products.

The challenge of globalization for the organic movement is the entry into the global industry without loosing its own identity. The increasing need for supply of organic products has led to a global marketplace, and with it, relationships in the market are continuously redefined; the system is subject to dynamic change caused by technology-driven progress. Suppliers of goods and services are no longer bound to local, regional or national markets. They can offer and reach everyone through the Internet, e-mail and mobile technologies. The customers of today receive all types of information with various levels of detail and depth. They can learn about the attributes, quality, price and availability of products, or exchange their experiences through the user community. Customers are now well informed, challenging and more demanding.

With the rise in 'buy local' campaigns, the issue of food miles, or the distance that food travels before it reaches the consumer, has become an important topic. As the demand for organic products increased, so has the amount of products sourced from distant places, thereby increasing the amount of food miles attached to organic products. This practice has been criticized as unsustainable and inefficient on a per calorie basis. Others argue that organic's carbon footprint is still better because it requires less energy from farm to plate overall.

It is indeed important to source locally whenever possible, and we must think of ways to reduce our carbon footprint. However, such moves could dramatically decrease the global conversion to organic while simultaneously reducing sourcing options and the variety of organic products available.

The impetus for revising the Organic Guarantee System comes from the need to respond to changes in the organic sector, particularly the rapid worldwide expansion of organic production and marketing, and the government regulation of organic agriculture. Beside this, the standards and regulations differentiate organic from conventional agriculture. The basic values and principles, and the connection between principles and regulations, is therefore of key concern to the organic market.

In 2007, the FAO recognized organic agriculture as a significant and singularly effective alternative to conventional, chemical-based agriculture. IFOAM will continue to participate at FAO conferences in 2008, actively engaging IFOAM members and representatives. IFOAM will guarantee that the organic voice is heard at critical policy discussions worldwide, including at the United Nations Conference on Biodiversity and Planet Diversity (the World Congress on the Future of Food and Agriculture) to be held in Bonn in May 2008.

As you can see, IFOAM has not lost the vision of its original founders and will continue to follow its mission n to lead, unite and assist the organic movement in its full diversity in the coming years.

ANNEX

In this section the results of the FiBL Survey 2008 are presented in full detail (except the results on certified wild collection, which are in the continent chapters). Should you notice errors please send the correct data to helga.willer@fibl.org; we will make these available at www.organic-world.net and in the 2009 edition of 'The World of Organic Agriculture.'

Table 36: Alphabetical list of countries: Organic land, share of organic land of total agricultural area, number of organic farms 2006

Data as of 2006 were not available for all countries.

Country	Organic agricultural land (ha)	Share of total agricultural land	Organic farms
Albania	1'000	0.1%	100
Algeria	1'550	No data	39 (2005)
Argentina	2'220'489	1.7%	1'486
Armenia	235	0.02%	35
Australia	12'294'290	2.8%	1'550
Austria	361'487	13%	20'162
Azerbaijan	20'779	0.4%	388
Belgium	29'308	2.1%	783
Belize (2000)	1'810	1.2%	
Benin	825	0.02%	1'132
Bhutan	243	0.04%	53
Bolivia	41'004	0.1%	11'743
Bosnia Herzegovina	726	0.03%	329
Brazil	880'000	0.3%	15'000 (2005)
Bulgaria	4'692	0.2%	218
Burkina Faso	4'038	0.04%	6'195
Cambodia	1'451	0.03%	3'628
Cameroon	531	0.01%	102
Canada	604'404	0.9%	3'571
Chad[1]		0.00%	36
Chile	9'464	0.06%	1'000
China	2'300'000	0.4%	1'600
Colombia	50'713	0.1%	4'500
Congo (Democr. Rep.)	8'788	0.04%	5'150
Costa Rica	10'711	0.4%	2'921
Croatia	6'204	0.2%	368
Cuba (2005)	15'443	0.2%	7'101
Cyprus	1'979	1.3%	305 (2005)
Czech Rep.	281'535	6.6%	963
Denmark	138'079	5.3%	2'794
Dominican Rep.	47'032	1.3%	4'638
Ecuador	50'475	0.6%	137
Egypt	14'165	0.4%	460
El Salvador	7'469	0.6%	1'811
Estonia	72'886	8.8%	1'173
Ethiopia	2'601	0.01%	784

---

[1] Chad has wild collection activities. The number of farms refers to these activities.

Country	Organic agricultural land (ha)	Share of total agricultural land	Organic farms
Fiji (2005)	100	0.02%	No data
Finland	144'558	6.4%	3'966
France	552'824	2.0%	11'640
Gambia	86	0.01%	No data
Georgia	247	0.01%	47
Germany	825'539	4.8%	17'557
Ghana	22'276	0.2%	3'000
Greece	302'256	7.6%	23'900
Guatemala (2005)	12'110	0.3%	2'830
Guyana (2003)	109	0.01%	28
Honduras	12'866	0.4%	1'813
Hong Kong (2005)	12		20
Hungary	122'765	2.9%	1'553 (2005)
Iceland	5'512	0.4%	27
India	528'171	0.3%	44'926
Indonesia	41'431	0.1%	23'608
Iran	15		2
Ireland	39'947	1%	1'104
Israel	4'058	0.7%	216
Italy	1'148'162	9.0%	45'115
Ivory Coast	13'311	0.07%	No data
Jamaica	437	0.09%	11
Japan	6'074	0.2%	2'258
Jordan	1'024	0.1%	25
Kazakhstan	2'393		
Kenya	3'307	0.01%	18'056
Korea, Republic of	8'559	0.5%	7'167
Kyrgyzstan	2'540	0.02%	392
Latvia	118'612 (2005)	7% (2005)	4'095
Lebanon	3'470	1.0%	213
Liechtenstein	1'027	29.1%	41
Lithuania	96'718	3.5%	1'811 (2005)
Luxemburg	3'630	2.8%	72
Macedonia	509	0.04%	101
Madagascar	9'456	0.03%	5'455
Malawi (2003)	325	0.01%	13
Malaysia	1'000	0.01%	50
Mali	2'330	0.01%	5'840
Malta	20	0.2%	10
Mauritius	175	0.2%	5
Mexico	404'118	0.4%	126'000
Moldova	11'405	0.5%	121 (2005)
Montenegro	25'051	4.8%	15
Morocco	4'216	0.01%	No data
Mozambique	728	0.00%	1'928
Nepal	7'762	0.2%	1'183
Netherlands	48'424	2.5%	1'448
New Zealand	63'883	0.4%	860
Nicaragua	60'000	0.9%	6'600
Niger	81		No data
Nigeria	3'042		No data
Niue	159	2%	61
Norway	44'624	4.3%	2'583
Pakistan	25'001	0.10%	28 (2004)
Palestine, Occupied Tr.	641	0.2%	303

Country	Organic agricultural land (ha)	Share of total agricultural land	Organic farms
Panama	5'267	0.2%	7 (2004)
Papua New Guinea	2'497	0.2%	4'558
Paraguay	17'705	0.07%	3'490
Peru	121'677	0.6%	31'530
Philippines	5'691	0.05%	No data
Poland	228'009	1.6%	9'187
Portugal	269'374	7.3%	1'696
Romania	107'582	0.8%	3'033
Russian Federation	3'192	0.00%	8
Rwanda	512	0.03%	20
Samoa	7'243	5.5%	213
Sao Tome and Prince	2'917	5.2%	1'291
Saudi Arabia (2005)	13'730	0.01%	3
Senegal	130	0.00%	1'020
Serbia	906	0.02%	35
Slovak Republic	121'461	5.8%	279
Slovenia	26'831	5.5%	1'953
Solomon Islands	3'628	3.1%	352
South Africa (2005)	50'000	0.05%	No data
Spain	926'390	3.7%	17'214
Sri Lanka	17'000	0.7%	4'216
Sweden	225'385	7.0%	2'380
Switzerland	125'596	11.8%	6'563
Syria	30'493	0.2%	3'256
Taiwan (March 2007)	1'746	0.2%	905
Tanzania	23'732	0.05%	22'301
Thailand	21'701	0.1%	2'498
Timor Leste	23'589	6.9%	No data
Togo	2'338	0.06%	5'101
Trinidad & Tobago (2005)	67	0.05%	1
Tunisia	154'793	1.6%	862
Turkey	100'275	0.4%	14'256
Uganda	88'439	0.7%	86'952
UK	604'571	3.8%	4'485
Ukraine	260'034	0.6%	80
Uruguay	930'965	6.1%	630
USA (2005)	1'620'351	0.5%	8'493
Vanuatu	8'996	6.1%	No data
Venezuela	15'712	0.07%	No data
Vietnam	21'867	0.2%	No data
Zambia	2'367	0.01%	9'524
Total	30'418'261	0.65%	718'744

Table 37: Organic agricultural land by country 2006, sorted by importance

Country	Organic agr. land (ha)	Country	Organic agr. land (ha)
Australia	12'294'290	Spain	926'390
China	2'300'000	Brazil	880'000
Argentina	2'220'489	Germany	825'539
USA (2005)	1'620'351	UK	604'571
Italy	1'148'162	Canada	604'404
Uruguay	930'965	France	552'824

235

Country	Organic agr. land (ha)
India	528'171
Mexico	404'118
Austria	361'487
Greece	302'256
Czech Rep.	281'535
Portugal	269'374
Ukraine	260'034
Poland	228'009
Sweden	225'385
Tunisia	154'793
Finland	144'558
Denmark	138'079
Switzerland	125'596
Hungary	122'765
Peru	121'677
Slovak Republic	121'461
Latvia (2005)	118'612
Romania	107'582
Turkey	100'275
Lithuania	96'718
Uganda	88'439
Estonia	72'886
New Zealand	63'883
Nicaragua	60'000
Colombia	50'713
Ecuador	50'475
South Africa (2005)	50'000
Netherlands	48'424
Dominican Rep.	47'032
Norway	44'624
Indonesia	41'431
Bolivia	41'004
Ireland	39'947
Syria	30'493
Belgium	29'308
Slovenia	26'831
Tanzania	23'732
Montenegro	25'051
Pakistan	25'001
Timor Leste	23'589
Ghana	22'276
Vietnam	21'867
Thailand	21'701
Azerbaijan	20'779
Paraguay	17'705
Sri Lanka	17'000
Venezuela	15'712
Cuba (2005)	15'443
Egypt	14'165
Saudi Arabia (2005)	13'730
Ivory Coast	13'311
Honduras	12'866
Guatemala (2005)	12'110
Moldova	11'405
Costa Rica	10'711

Country	Organic agr. land (ha)
Chile	9'464
Madagascar	9'456
Vanuatu	8'996
Congo (Democr. Rep.)	8'788
Korea, Republic of	8'559
Nepal	7'762
El Salvador	7'469
Samoa	7'243
Croatia	6'204
Japan	6'074
Philippines	5'691
Iceland	5'512
Panama	5'267
Bulgaria	4'692
Morocco	4'216
Israel	4'058
Burkina Faso	4'038
Luxemburg	3'630
Solomon Islands	3'628
Lebanon	3'470
Kenya	3'307
Russian Federation, European Part	3'192
Nigeria	3'042
Sao Tome and Prince	2'917
Ethiopia	2'601
Kyrgyzstan	2'540
Papua New Guinea	2'497
Kazakhstan	2'393
Zambia	2'367
Togo	2'338
Mali	2'330
Cyprus	1'979
Belize (2000)	1'810
Taiwan (March 2007)	1'746
Algeria	1'550
Cambodia	1'451
Liechtenstein	1'027
Jordan	1'024
Albania	1'000
Malaysia	1'000
Serbia	906
Benin	825
Mozambique	728
Bosnia Herzegovina	726
Palestine, Occupied Tr.	641
Cameroon	531
Rwanda	512
Macedonia	509
Jamaica	437
Malawi (2003)	325
Georgia	247
Bhutan	243
Armenia	235
Mauritius	175

Country	Organic agr. land (ha)
Niue	159
Senegal	130
Guyana (2003)	109
Fiji (2005)	100
Gambia	86

Country	Organic agr. land (ha)
Niger	81
Trinidad & Tobago (2005)	67
Malta	20
Iran	15
Hong Kong	12

Source: FiBL Survey 2008

Table 38: Share of organic agricultural land by country, sorted by importance

Only countries with a share higher than 0.01 percent.

Country	Share of agr. land
Liechtenstein	29.1%
Austria	13.0%
Switzerland	11.8%
Italy	9.0%
Estonia	8.8%
Greece	7.6%
Portugal	7.3%
Sweden	7.1%
Latvia (2005)	7.0%
Timor Leste	6.9%
Czech Rep.	6.6%
Finland	6.4%
Vanuatu	6.1%
Uruguay	6.1%
Slovak Republic	5.8%
Samoa	5.5%
Slovenia	5.5%
Denmark	5.3%
Sao Tome and Prince	5.2%
Germany	4.8%
Montenegro	4.8%
Norway	4.3%
UK	3.8%
Spain	3.7%
Lithuania	3.5%
Solomon Islands	3.1%
Hungary	2.9%
Luxemburg	2.8%
Australia	2.8%
Netherlands	2.5%
Belgium	2.1%
France	2.0%
Niue	2.0%
Argentina	1.7%
Tunisia	1.6%
Poland	1.5%
Cyprus	1.3%
Dominican Rep.	1.3%
Belize (2000)	1.2%
Lebanon	1.0%
Ireland	0.9%
Canada	0.9%
Nicaragua	0.9%

Country	Share of agr. land
Romania	0.8%
Sri Lanka	0.7%
Israel	0.7%
Uganda	0.7%
Ukraine	0.6%
Ecuador	0.6%
El Salvador	0.6%
Peru	0.6%
USA (2005)	0.5%
Moldova	0.5%
Korea, Republic of	0.4%
Azerbaijan	0.4%
Honduras	0.4%
China	0.4%
Egypt	0.4%
Turkey	0.4%
Mexico	0.4%
Costa Rica	0.4%
New Zealand	0.4%
Iceland	0.4%
Brazil	0.3%
India	0.3%
Guatemala (2005)	0.3%
Papua New Guinea	0.2%
Panama	0.2%
Cuba	0.2%
Vietnam	0.2%
Syria	0.2%
Taiwan (March 2007)	0.2%
Croatia	0.2%
Malta	0.2%
Palestine, Occupied Tr.	0.2%
Nepal	0.2%
Bulgaria	0.2%
Japan	0.2%
Mauritius	0.2%
Ghana	0.2%
Thailand	0.1%
Bolivia	0.1%
Colombia	0.1%
Pakistan	0.1%
Indonesia	0.1%
Jordan	0.1%

Country	Share of agr. land
Albania	0.1%
Jamaica	0.1%
Venezuela	0.1%
Paraguay	0.1%
Ivory Coast	0.1%
Togo	0.1%
Chile	0.1%
Tanzania	0.1%
Trinidad & Tobago (2005)	0.1%
South Africa (2005)	0.1%
Philippines	0.05%
Bhutan	0.04%
Macedonia	0.04%
Congo (Democr. Rep.)	0.04%
Burkina Faso	0.04%
Madagascar	0.03%
Bosnia Herzegovina	0.03%
Cambodia	0.03%

Source: FiBL Survey 2008

Country	Share of agr. land
Rwanda	0.03%
Benin	0.02%
Kyrgyzstan	0.02%
Fiji (2005)	0.02%
Armenia	0.02%
Serbia	0.02%
Morocco	0.01%
Malaysia	0.01%
Kenya	0.01%
Gambia	0.01%
Georgia	0.01%
Ethiopia	0.01%
Saudi Arabia (2005)	0.01%
Malawi (2003)	0.01%
Mali	0.01%
Zambia	0.01%
Guyana (2003)	0.01%
Cameroon	0.01%

Table 39: Organic land use by country 2006

Please note: In some cases, the added up land use data resulted in a higher figure than the figure provided for the total organic area. This discrepancy is probably due to the fact that some of the organic land is used for two main crops. In such cases, minus values were added.

Country	Main use	Crop category	Crop	Year	Organic agr. area (ha)
Albania	No information	No information	No information	2006	995
	Permanent crops	Grapes	Grapes	2005	5
Algeria	No information	No information	No information	2006	1'550
Argentina	Arable land	Arable crops, no details	Arable crops, no details	2006	35'383
		Medicinal & aromatic plants	Medicinal & aromatic plants	2006	205
		Vegetables	Vegetables, no details	2006	2'482
	Permanent crops	Fruit and nuts	Fruit and nuts, no details	2006	3'906
	Permanent grassland	Pastures and meadows	Pastures and meadows	2006	2'164'200
	Cropland, other/no details	Cropland, no details	Cropland, no details	2006	14'314
Armenia	Arable land	Cereals	Barley	2006	5
			Grain Maize	2006	5
			Other cereals	2006	1
			Wheat	2006	11
		Green fodder from arable land	Lucerne	2006	11
		Protein crops	Feed legumes	2006	5
		Root crops	Fodder beet	2006	0.1
		Seeds and seedlings	Seeds and seedlings	2006	3
		Vegetables	Vegetables, no details	2006	12
	No information	No information	No information	2006	11
	Permanent crops	Fruit and nuts	Apples	2006	1
			Apricots	2006	116
			Cherries	2006	12

Country	Main use	Crop category	Crop	Year	Organic agr. area (ha)
			Cherries, sour	2006	2
			Fruit and nuts, no details	2006	4
			Peaches	2006	18
			Plums	2006	4
			Raspberries	2006	6
		Grapes	Grapes	2006	8
Australia	Permanent grassland	Permanent grassland, no details	Permanent grassland, no details	2006	11'925'461
	Cropland, other/no details	Cropland, no details	Cropland, no details	2006	368'829
Austria	Arable land	Cereals	Barley	2006	9'864
			Grain Maize	2006	6'024
			Oats	2006	10'178
			Other cereals	2006	2'218
			Rye and meslin	2006	5'599
			Triticale	2006	2'619
			Wheat	2006	27'343
		Fallow land as part of crop rotation	Fallow land as part of crop rotation	2006	5'555
		Flowers and ornamental plants	Flowers and ornamental plants	2006	5
		Green fodder from arable land	Field fodder growing	2006	46'203
		Industrial crops	Hops	2006	18
			Other Industrial crops	2006	285
		Medicinal & aromatic plants	Other Medicinal & aromatic plants	2006	969
		Oilseeds	Flax/linseed	2006	411
			Other oilseeds	2006	2'528
			Rape and Turnip rape	2006	191
			Soy beans	2006	2'623
			Sunflower seed	2006	1'600
		Protein crops	Dried pulses	2006	12'606
		Root crops	Fodder beet	2006	9
			Potatoes	2006	2'426
			Sugar beets	2006	334
		Seeds and seedlings	Seeds and seedlings	2006	380
		Vegetables	Fresh Vegetables	2006	1'497
	No information	No information	No information	2006	1'505
	Permanent crops	Fruit and nuts	Fruit, no details	2006	1'453
		Grapes	Grapes	2006	1'766
	Permanent grassland	Pastures and meadows	Pastures and meadows	2006	186'499
		Rough Grazing	Rough Grazing	2006	28'779
Azerbaijan	Arable land	Cereals	Cereals, no details	2006	1'852
		Flowers and ornamental plants	Flowers and ornamental plants	2006	2
		Green fodder from arable land	Other field fodder growing	2006	200
		Oilseeds	Sunflower seed	2006	73
		Protein crops	Other protein crops	2006	4
		Root crops	Potatoes	2006	215
			Sugar beets	2006	12
		Textile fibers	Cotton	2006	150
		Vegetables	Other vegetables	2006	322
			Strawberries	2006	28
	No information	No information	No information	2006	

239

Country	Main use	Crop category	Crop	Year	Organic agr. area (ha)
	Permanent crops	Fruit and nuts	Apples	2006	80
			Apricots	2006	58
			Blackberries	2006	8
			Cherries	2006	27
			Cherries, sour	2006	34
			Chestnuts	2006	55
			Hazelnut	2006	120
			Nuts, no details	2006	180
			Peaches	2006	45
			Pears	2006	72
			Plums	2006	40
			Pomegranate	2006	85
			Raspberries	2006	20
		Grapes	Grapes	2006	72
		Olives	Olives	2006	5
		Tea	Tea	2006	20
	Permanent grassland	Permanent grassland, no details	Permanent grassland, no details	2006	17'000
Belgium	Arable land	Arable crops, no details	Arable crops, no details	2006	4
		Cereals	Barley	2006	393
			Grain Maize	2006	116
			Oats	2006	215
			Other cereals	2006	1'349
			Rye and meslin	2006	38
			Wheat	2006	952
		Fallow land as part of crop rotation	Fallow land as part of crop rotation	2006	227
		Flowers and ornamental plants	Flowers and ornamental plants	2006	7
		Green fodder from arable land	Other field fodder growing	2006	5'292
		Industrial crops	Hops	2006	8
			Other Industrial crops	2006	14
			Rape and Turnip rape	2006	69
			Sunflower seed	2006	1
		Medicinal & aromatic plants	Other Medicinal & aromatic plants	2006	3
		Protein crops	Dried pulses	2006	264
		Root crops	Beets	2006	8
			Fodder roots and brassicas	2006	28
			Potatoes	2006	223
			Sugar beets	2006	8
		Seeds and seedlings	Seeds and seedlings	2006	4
		Vegetables	Other vegetables	2006	524
	Other	Unutilized land	Fallow land	2006	204
	Permanent crops	Fruit and nuts	Fruit, no details	2006	463
		Permanent crops, no details	Permanent crops, no details	2006	3
	Permanent grassland	Permanent grassland, no details	Permanent grassland, no details	2006	18'891
Belize	No information	No information	No information	2000	1'810
Benin	Arable land	Textile fibers	Cotton	2006	825
	No information	No information	No information	2006	
Bhutan	No information	No information	No information	2006	
	Permanent crops	Medicinal & aromatic plants	Lemongrass	2006	182

Country	Main use	Crop category	Crop	Year	Organic agr. area (ha)
	Permanent grassland	Permanent grassland, no details	Permanent grassland, no details	2006	61
Bolivia	No information	No information	No information	2006	41'004
Bosnia Herzegovina	No information	No information	No information	2006	726
Brazil	Arable land	Arable crops, no details	Arable crops, no details	2005	170'000
	No information	No information	No information	2006	38'000
	Permanent grassland	Permanent grassland, no details	Permanent grassland, no details	2005	672'000
Bulgaria	Arable land	Cereals	Cereals, no details	2006	191
		Green fodder from arable land	Field fodder growing	2006	30
		Industrial crops	Other Industrial crops	2006	146
		Medicinal & aromatic plants	Medicinal & aromatic plants	2006	709
		Protein crops	Dried pulses	2006	2
		Root crops	Potatoes	2006	14
		Vegetables	Vegetables, no details	2006	96
	No information	No information	No information	2006	2
	Other	Fallow land	Fallow land	2006	1'261
	Permanent crops	Fruit and nuts	Fruit, no details	2006	1'550
		Grapes	Grapes	2006	228
	Permanent grassland	Permanent grassland, no details	Permanent grassland, no details	2006	465
Burkina Faso	Arable land	Cereals	Fonio	2006	157
		Oilseeds	Peanuts	2006	443
			Sesame seeds	2006	416
		Textile fibers	Cotton	2006	834
		Vegetables	Peas	2006	50
	No information	No information	No information	2006	
	Permanent crops	Other permanent crops	Hibiscus	2006	738
		Tropical fruit and nuts	Cashew nuts	2006	75
			Mano	2006	1'326
Cambodia	Arable land	Cereals	Rice	2006	1'451
Cameroon	Arable land	Fallow land as part of crop rotation	Fallow land as part of crop rotation	2006	50
		Medicinal & aromatic plants	Ginger	2006	20
		Root crops	Potatoes	2006	50
	No information	No information	No information	2006	
	Other	Fallow land	Fallow land	2006	150
	Permanent crops	Cocoa	Cocoa	2006	0
		Sugarcane	Sugarcane	2006	5
		Tropical fruit and nuts	Avocado	2006	30
			Banana	2006	50
			Coconuts	2006	0
			Guineo	2006	31
			Mango	2006	75
			Papaya	2006	30
			Passion fruit	2006	5
			Pineapples	2006	35
Canada	Arable land	Cereals	Barley	2005	15'493
			Grain Maize	2005	2'280
			Kamut	2005	3'462
			Oats	2005	37'231

Country	Main use	Crop category	Crop	Year	Organic agr. area (ha)
			Other cereals	2005	12'038
			Rye and meslin and meslin	2005	7'196
			Triticale	2005	636
			Wheat	2005	75'816
		Fallow land as part of crop rotation	Fallow land as part of crop rotation	2005	53'480
		Flowers and ornamental plants	Other Flowers and ornamental plants	2005	24
		Green fodder from arable land	Lucerne	2005	8'774
			Other field fodder growing	2005	79'576
		Medicinal & aromatic plants	Herbs for essential oil	2005	389
			Other Medicinal & aromatic plants	2005	473
		Oilseeds	Rape seeds	2005	857
		Protein crops	Dried pulses	2005	112
			Lentils	2005	14'942
			Peas	2005	12'293
			Soy	2005	8'062
		Root crops	Potatoes	2005	497
		Textile fibers	Flax	2005	32'754
		Vegetables	Greenhouse cultivation	2005	32
			Other vegetables	2005	1'571
			Strawberries	2005	13
	No information	No information	No information	2005	167'588
	Permanent crops	Fruit and nuts	Apples	2005	606
			Apricots	2005	8
			Blueberries	2005	90
			Cherries	2005	13
			Chestnuts	2005	2
			Fruit and nuts, no details	2005	127
			Hazelnut	2005	39
			Nuts, no details	2005	18
			Peaches	2005	18
			Pears	2005	27
			Plums	2005	6
			Raspberry	2005	17
			Walnuts	2005	4
		Grapes	Grapes	2005	69
				2006	
	Permanent grassland	Pastures and meadows	Pastures and meadows	2005	67'771
Chile	Arable land	Arable crops, no details	Arable crops, no details	2006	291
		Green fodder from arable land	Green fodder from arable land	2006	1'085
		Medicinal & aromatic plants	Medicinal & aromatic plants	2006	450
		Vegetables	Vegetables, no details	2006	263
	Other	Forest	Forest	2006	690
	Permanent crops	Fruit and nuts	Fruit, no details	2006	3'227
		Grapes	Grapes	2006	2'477
	Cropland, other/no details	Cropland, no details	Cropland, no details	2006	981
China	No information	No information	No information	2006	610'000

242

Country	Main use	Crop category	Crop	Year	Organic agr. area (ha)
	Permanent grassland	Permanent grassland, no details	Permanent grassland, no details	2006	692'000
	Cropland, other/no details	Cropland, no details	Cropland, no details	2006	998'000
Colombia				2006	50'713
Congo (Democr. Rep.)	No information	No information	No information	2006	8'788
Costa Rica	Arable land	Cereals	Rice	2006	48
		Medicinal & aromatic plants	Sabila	2006	72
	No information	No information	No information	2006	127
	Permanent crops	Citrus fruit	oranges	2006	1'143
		Cocoa	Cocoa	2006	2'382
		Fruit and nuts	Noni	2006	51
		Sugarcane	Sugarcane	2006	324
		Tropical fruit and nuts	Banana	2006	3'938
			Blackberries	2006	700
			Coffee	2006	1'524
			Macadamia	2006	46
			Mango	2006	25
			Pineapples	2006	332
Croatia	Arable land	Arable crops, no details	Arable crops, no details	2006	1'244
		Cereals	Cereals, no details	2006	1'713
		Medicinal & aromatic plants	Medicinal & aromatic plants	2006	159
		Vegetables	Vegetables, no details	2006	35
	No information	No information	No information	2006	
	Other	Fallow land	Fallow land	2006	101
		Forest	Forest	2006	59
	Permanent crops	Fruit and nuts	Fruit and nuts, no details	2006	204
		Grapes	Grapes	2006	32
		Olives	Olives	2006	37
	Permanent grassland	Pastures and meadows	Pastures and meadows	2006	2'620
Cuba	Arable land	Other arable crops	self sufficiency	2005	5'000
	Permanent crops	Citrus fruit	Other citrus fruit	2005	1'735
		Cocoa	Cocoa	2005	1'369
		Coffee	Coffee	2005	490
		Sugarcane	Sugarcane	2005	5'662
		Tropical fruit and nuts	Coconuts	2005	1'056
			Mango	2005	131
Cyprus	Arable land	Cereals	Barley	2006	306
			Oats	2006	27
			Wheat, durum	2006	23
		Medicinal & aromatic plants	Other Medicinal & aromatic plants	2006	33
		Protein crops	Dried pulses	2006	543
		Root crops	Potatoes	2006	7
		Vegetables	Vegetables, no details	2006	13
	No information	No information	No information	2006	
	Permanent crops	Citrus fruit	Other citrus fruit	2006	24
		Fruit and nuts	Berries, no details	2006	7
			Fruit and nuts, no details	2006	95
			Nuts, no details	2006	47
		Grapes	Grapes	2006	119

Country	Main use	Crop category	Crop	Year	Organic agr. area (ha)
		Olives	Olives	2006	665
		Permanent crops, no details	Permanent crops, no details	2006	70
Czech Rep.	Arable land	Arable crops, no details	Arable crops, no details	2005	20'766
	No information	No information	No information	2005	26'553
	Other	Unutilized land	Unutilized land	2005	23'440
	Permanent crops	Fruit and nuts	Fruit, no details	2005	772
		Grapes	Grapes	2005	48
	Permanent grassland	Permanent grassland, no details	Permanent grassland, no details	2005	209'956
Denmark	Arable land	Arable crops, no details	Arable crops, no details	2006	43
		Cereals	Barley	2006	10'566
			Grain Maize	2006	15
			Oats	2006	9'098
			Other cereals	2006	10'245
			Rye and meslin	2006	2'813
			Wheat	2006	6'929
		Flowers and ornamental plants	Flowers and ornamental plants	2006	4
		Green fodder from arable land	Other field fodder growing	2006	63'410
		Industrial crops	Other Industrial crops	2006	228
			Rape and Turnip rape	2006	4'447
		Medicinal & aromatic plants	Medicinal & aromatic plants	2006	30
		Protein crops	Dried pulses	2006	2'039
		Root crops	Fodder beet	2006	23
			Potatoes	2006	968
		Seeds and seedlings	Seeds and seedlings	2006	1'777
		Vegetables	Other vegetables	2006	1'166
	Other	Unutilized land	Unutilized land	2006	1'797
	Permanent crops	Fruit and nuts	Fruit, no details	2006	264
			Nuts, no details	2006	0
		Permanent crops, no details	Permanent crops, no details	2006	2'999
	Permanent grassland	Permanent grassland, no details	Permanent grassland, no details	2006	19'218
Dominican Rep.	Arable land	Vegetables	Vegetables, no details	2006	18
	No information	No information	No information	2006	2'017
	Permanent crops	Citrus fruit	Other citrus fruit	2005	557
		Cocoa	Cocoa	2006	30'624
		Coffee	Coffee	2006	8'135
		Other permanent crops	Yucca	2006	600
		Tropical fruit and nuts	Avocado	2006	110
			Banana	2006	2'637
			Coconuts	2006	1'907
			Macadamia	2006	78
			Mango	2006	62
	Cropland, other/no details	Cropland, no details	Cropland, no details	2005	288
Ecuador	Arable land	Cereals	Barley	2006	45
			Other cereals	2006	25
			Quinoa	2006	316
			Rice	2006	20
		Flowers and ornamen-	Cartucho	2006	7

Country	Main use	Crop category	Crop	Year	Organic agr. area (ha)
		tal plants			
		Medicinal & aromatic plants	Other Medicinal & aromatic plants	2006	52
		Oilseeds	Mani	2006	31
			Peanuts	2006	31
		Other arable crops	Panela	2006	74
		Vegetables	Vegetables, no details	2006	49
	No information	No information	No information	2006	1'297
	Other	Aquaculture	Aquaculture, no details	2006	3'210
	Permanent crops	Citrus fruit	Citrus fruit, no details	2006	2'023
		Cocoa	Cocoa	2006	20'018
		Coffee	Coffee	2006	4'351
		Fruit and nuts	Blackberries	2006	22
		Sugarcane	Sugarcane	2006	268
		Tropical fruit and nuts	Araza	2006	41
			Banana	2006	14'810
			Mango	2006	455
			Orito	2006	3'180
			Papaya	2006	7
			Pineapples	2006	144
Egypt	Cropland, other/no details	Cropland, no details	Cropland, no details	2006	14'165
El Salvador	Arable land	Medicinal & aromatic plants	Indigofera suffruticosa	2006	7
		Oilseeds	Sesame seeds	2006	916
	Permanent crops	Coffee	Coffee	2006	3'325
		Tropical fruit and nuts	Cashew nuts	2006	117
			Coconuts	2006	895
	Cropland, other/no details	Cropland, no details	Cropland, no details	2006	2'210
Estonia	Arable land	Cereals	Barley	2006	2'277
			Oats	2006	3'396
			Other cereals	2006	851
			Rye and meslin	2006	652
			Wheat and spelt	2006	461
			Wheat, durum	2006	883
		Fallow land as part of crop rotation	Fallow land as part of crop rotation	2006	1'660
		Green fodder from arable land	Field fodder growing	2006	48'276
		Industrial crops	Other Industrial crops	2006	29
		Medicinal & aromatic plants	Medicinal & aromatic plants	2006	151
		Oilseeds	Rape and Turnip rape	2006	283
		Protein crops	Dried pulses	2006	149
		Root crops	Fodder beet	2006	6
			Potatoes	2006	241
		Vegetables	Vegetables, no details	2006	96
	Other	Fallow land	Fallow land	2006	497
	Permanent crops	Fruit and nuts	Fruit, no details	2006	1'146
	Permanent grassland	Permanent grassland, no details	Permanent grassland, no details	2006	11'832
Ethiopia	No information	No information	No information	2006	2'601
Fiji	Permanent crops	Tropical fruit and nuts	Coconuts	2005	100
Finland	Arable land	Cereals	Barley	2006	4'819
			Buckwheat	2006	114
			Mixed grains	2006	2'805

245

Country	Main use	Crop category	Crop	Year	Organic agr. area (ha)
			Oats	2006	19'283
			Other cereals	2006	2'858
			Rye, spring	2006	1'427
			Rye, winter	2006	3'523
			Spelt	2006	454
			Wheat, spring	2006	4'476
			Wheat, winter	2006	430
		Green fodder from arable land	Other field fodder growing	2006	264
		Medicinal & aromatic plants	Medicinal & aromatic plants	2006	17
		Oilseeds	Linseed	2006	141
			Spring turnip rape	2006	2'684
			Sunflower seed	2006	7
		Protein crops	Broad beans	2006	415
			Fodder pea	2006	460
			Pea	2006	766
		Root crops	Potatoes	2006	356
		Vegetables	Greenhouse cultivation	2006	8
			Vegetables, no details	2006	197
	No information	No information	No information	2006	17'013
	Other	Fallow land	Fallow land	2006	6'541
	Permanent crops	Fruit and nuts	Berries, no details	2006	723
	Permanent grassland	Cultivated grassland	Cultivated grassland	2006	74'778
France	Arable land	Cereals	Barley	2006	9'098
			Buckwheat	2006	2'392
			Grain Maize	2006	7'186
			Millet	2006	533
			Oats	2006	4'117
			Other cereals	2006	13'205
			Rice	2006	940
			Rye and meslin and meslin	2006	2'041
			Spelt	2006	2'455
			Triticale	2006	9'341
			Wheat, durum	2006	2'407
			Wheat, soft	2006	30'146
		Green fodder from arable land	Field fodder growing	2006	122'512
		Medicinal & aromatic plants	Medicinal & aromatic plants	2006	2'438
		Oilseeds	Linseed	2006	405
			Other oilseeds	2006	60
			Rape and Turnip rape	2006	1'892
			Soy	2006	6'614
			Sunflower seed	2006	9'737
		Protein crops	Protein crops	2006	11'151
		Vegetables	Vegetables, no details	2006	8'767
	No information	No information	No information	2006	
	Permanent crops	Fruit and nuts	Fruit, no details	2006	9'179
		Grapes	Grapes	2006	18'808
	Permanent grassland	Permanent grassland, no details	Permanent grassland, no details	2006	219'763
	Cropland, other/no details	Cropland, no details	Cropland, no details	2006	57'637
Gambia	Arable land	Arable crops, no details	Arable crops, no details	2006	-51

Country	Main use	Crop category	Crop	Year	Organic agr. area (ha)
		Protein crops	Beans	2006	70
		Vegetables	Okra	2006	5
			Pepper	2006	30
	No information	No information	No information	2006	
	Permanent crops	Citrus fruit	Other citrus fruit	2006	1
		Tropical fruit and nuts	Mangos	2006	31
Georgia	Arable land	Flowers and ornamental plants	Roses	2006	32
	No information	No information	No information	2006	113
	Permanent crops	Grapes	Grapes	2006	102
Germany	Arable land	Cereals	Barley	2006	20'500
			Grain Maize	2006	4'500
			Oats	2006	18'800
			Other cereals	2006	5'200
			Rye and meslin	2006	49'000
			Spelt	2006	17'000
			Triticale	2006	18'000
			Wheat, summer	2006	7'500
			Wheat, winter	2006	38'500
		Fallow land as part of crop rotation	Fallow land as part of crop rotation	2006	14'000
		Flowers and ornamental plants	Flowers and ornamental plants	2006	175
		Green fodder from arable land	Field fodder growing	2006	122'000
		Medicinal & aromatic plants	Medicinal & aromatic plants	2006	700
		Oilseeds	Other oilseeds	2006	2'200
			Rape and Turnip rape	2006	2'800
			Soy	2006	600
			Sunflower seed	2006	2'200
		Other arable crops	Hemp	2006	250
		Protein crops	Protein crops	2006	28'000
		Root crops	Other root crops	2006	0
			Potatoes	2006	7'500
			Sugar beets	2006	1'000
		Vegetables	Vegetables, no details	2006	8'900
	No information	No information	No information	2006	6'284
	Permanent crops	Fruit and nuts	Berries, no details	2006	600
			Extensive fruit production	2006	11'000
			Fruit, no details	2006	1'900
			Pomefruits	2006	2'700
			Stone fruit	2006	400
		Grapes	Grapes	2006	2'700
		Industrial crops	Hops	2006	80
		Nurseries	Nurseries	2006	550
	Permanent grassland	Permanent grassland, no details	Permanent grassland, no details	2006	430'000
Ghana	Permanent crops	Citrus fruit	Citrus fruit, no details	2006	3'760
		Cocoa	Cocoa	2006	400
		Other permanent crops	Oil palm	2006	14'524
		Tropical fruit and nuts	Banana	2006	60
			Mango	2006	1'912
			Papaya	2006	470
			Pineapples	2006	1'116

Country	Main use	Crop category	Crop	Year	Organic agr. area (ha)
	Cropland, other/no details	Cropland, no details	Cropland, no details	2006	34
Greece	Arable land	Cereals	Barley	2006	5'048
			Grain Maize	2006	4'237
			Oats	2006	6'074
			Other cereals	2006	497
			Rice	2006	203
			Rye and meslin and meslin	2006	1'623
			Triticale	2006	1'255
			Wheat, durum	2006	3'000
			Wheat, soft	2006	27'035
		Fallow land as part of crop rotation	Fallow land as part of crop rotation	2006	13'998
		Flowers and ornamental plants	Other Flowers and ornamental plants	2006	4
		Green fodder from arable land	Green fodder from arable land, no details	2006	29'044
			Maize for silage	2006	600
		Industrial crops	Other Industrial crops	2006	18
			Tobacco	2006	29
		Medicinal & aromatic plants	Other Medicinal & aromatic plants	2006	821
		Oilseeds	Other oilseeds	2006	26
			Rape and Turnip rape	2006	6
			Soy	2006	12
			Sunflower seed	2006	1'197
		Protein crops	Beans	2006	24
			Other protein crops	2006	1'176
			Peas	2006	400
		Root crops	Potatoes	2006	103
			Sugar beets	2006	67
		Seeds and seedlings	Seeds and seedlings	2006	4
		Textile fibers	Cotton	2006	2'102
		Vegetables	Other vegetables	2006	1'362
	No information	No information	No information	2006	3
	Permanent crops	Citrus fruit	Grapefruit	2006	15
			Lemon	2006	212
			oranges	2006	2'019
			Other citrus fruit	2006	324
		Fruit and nuts	Apples	2006	193
			Apricots	2006	87
			Berries, no details	2006	2
			Cherries	2006	6
			Fruit, no details	2006	462
			Nuts, no details	2006	1'611
			Other stone fruit	2006	55
			Peaches	2006	55
			Pears	2006	153
			Plums	2006	22
		Grapes	Grapes	2006	
			Grapes for wine	2006	3'887
			Table grapes	2006	716
		Olives	Olives for eating	2006	12'244
			Olives for oil	2006	47'755
		Permanent crops, no details	Permanent crops, no details	2006	283

248

Country	Main use	Crop category	Crop	Year	Organic agr. area (ha)
	Permanent grassland	Permanent grassland, no details	Permanent grassland, no details	2006	117
		Rough Grazing	Rough Grazing	2006	132'069
Guatemala	Arable land	Cereals	Amaranth	2005	2
		Medicinal & aromatic plants	Other Medicinal & aromatic plants	2005	4
		Vegetables	Other vegetables	2005	8
	No information	No information	No information	2005	183
	Permanent crops	Citrus fruit	Other citrus fruit	2005	125
		Cocoa	Cocoa	2005	42
		Coffee	Coffee	2005	9'870
		Fruit and nuts	Blueberries	2005	5
		Medicinal & aromatic plants	Cardamon	2005	1'594
			Vanilla	2005	1
		Other permanent crops	Yucca	2005	3
		Tea	Tea	2005	26
		Tropical fruit and nuts	Banana	2005	51
			Macadamia	2005	163
			Mango	2005	25
			Papaya	2005	11
Guyana	No information	No information	No information	2003	109
Honduras	No information	No information	No information	2006	361
	Permanent crops	Citrus fruit	Lemon	2006	56
		Coffee	Coffee	2006	11'732
		Tropical fruit and nuts	Banana	2006	35
			Cashew nuts	2006	563
			Mangos	2006	111
			Pineapples	2006	9
Hong Kong	Arable land	Vegetables	Vegetables, no details	2005	12
Hungary	Arable land	Arable crops, no details	Arable crops, no details	2006	23'581
		Cereals	Cereals, no details	2006	21'600
	No information	No information	No information	2006	5'974
	Permanent crops	Fruit and nuts	Fruit, no details	2006	1'727
		Grapes	Grapes	2006	594
		Permanent crops, no details	Permanent crops, no details	2006	697
	Permanent grassland	Permanent grassland, no details	Permanent grassland, no details	2006	63'077
	Cropland, other/no details	Cropland, no details	Cropland, no details	2006	5'515
Iceland	No information	No information	No information	2006	5'512
India	No information	No information	No information	2006	528'171
Indonesia	Arable land	Arable crops, no details	Arable crops, no details	2006	23
	No information	No information	No information	2006	11'596
	Other	Aquaculture	Shrimp	2006	1'012
	Permanent crops	Coffee	Coffee	2006	25'300
			Coffee and spices	2006	3'000
		Tropical fruit and nuts	Cashew nuts	2006	500
Iran	Arable land	Medicinal & aromatic plants	Saffron	2006	11
	Permanent crops	Fruit and nuts	Figs	2006	4
Ireland	Arable land	Arable crops, no details	Arable crops, GhanaN-no details	2006	2'500
		Cereals	Cereals, no details	2006	800

Country	Main use	Crop category	Crop	Year	Organic agr. area (ha)
		Vegetables	Other vegetables	2006	400
	No information	No information	No information	2006	247
	Permanent grassland	Permanent grassland, no details	Permanent grassland, no details	2006	36'000
Israel	Arable land	Arable crops, no details	Arable crops, no details	2006	2'941
		Vegetables	Vegetables, protected crops	2006	258
	No information	No information	No information	2006	-100
	Permanent crops	Citrus fruit	Citrus fruit, no details	2006	399
		Fruit and nuts	Fruit, no details	2006	460
		Grapes	Grapes	2006	100
Italy	Arable land	Arable crops, no details	Arable crops, no details	2006	9'410
		Cereals	Barley	2006	32'834
			Oats	2006	37'693
			Other cereals	2006	15'294
			Rice	2006	13'670
			Rye and meslin and meslin	2006	1'315
			Wheat, durum	2006	117'686
			Wheat, soft	2006	20'599
		Fallow land as part of crop rotation	Fallow land as part of crop rotation	2006	27'006
		Green fodder from arable land	Other field fodder growing	2006	297'441
		Industrial crops	Other fiber plants	2006	219
			Other Industrial crops	2006	1'437
			Tobacco	2006	121
		Medicinal & aromatic plants	Medicinal & aromatic plants	2006	2'878
		Oilseeds	Oilseeds	2006	18'703
		Protein crops	Protein crops	2006	8'422
		Root crops	Root crops	2006	2'317
		Textile fibers	Cotton	2006	4
		Vegetables	Vegetables, no details	2006	39'696
	No information	No information	No information	2006	
	Other	Unutilized land	Unutilized land	2006	14'630
	Permanent crops	Citrus fruit	Grapefruit	2006	412
			Lemon	2006	3'482
			oranges	2006	10'979
			Other citrus fruit	2006	4'676
		Fruit and nuts	Apples	2006	2'873
			Apricots	2006	2'796
			Berries, no details	2006	684
			Cherries	2006	2'907
			Fruit, no details	2006	26'273
			Other stone fruit	2006	4'997
			Peaches	2006	2'963
			Pears	2006	1'420
			Plums	2006	759
		Grapes	Grapes	2006	37'693
		Olives	Olives	2006	107'233
		Permanent crops, no details	Permanent crops, no details	2006	15'387
	Permanent grassland	Permanent grassland, no details	Permanent grassland, no details	2006	261'253
Ivory Coast	No information	No information	No information	2006	13'311

Country	Main use	Crop category	Crop	Year	Organic agr. area (ha)
Jamaica	Arable land	Seeds and seedlings	Seedlings	2006	1
		Vegetables	Cabbage	2006	2
			Other vegetables	2006	4
			Pimento	2006	60
			Sorrel	2006	3
	No information	No information	No information	2006	
	Permanent crops	Citrus fruit	Citrus fruit, no details	2006	3
		Cocoa	Cocoa	2006	30
		Coffee	Coffee associated with other crops	2006	2
		Permanent crops, no details	Permanent crops, no details	2006	4
		Tropical fruit and nuts	Banana	2006	6
			Banana associated with other crops	2006	1
			Coconuts	2006	160
			Pineapples	2006	1
	Permanent grassland	Pastures and meadows	Pastures and meadows	2006	160
Japan	No information	No information	No information	2006	6'074
Jordan	No information	No information	No information	2006	1'024
Ka-zakhstan	No information	No information	No information	2006	2'393
Kenya	Arable land	Arable crops, no details	Arable crops, no details	2006	42
		Medicinal & aromatic plants	Borage	2006	4
			Other Medicinal & aromatic plants	2006	254
		Vegetables	Chilies	2006	171
			Onions	2006	24
			Other vegetables	2006	96
			Strawberries	2006	20
	No information	No information	No information	2006	190
	Permanent crops	Citrus fruit	Grapefruit	2006	201
		Coffee	Coffee	2006	156
		Permanent crops, no details	Permanent crops, no details	2006	1'817
		Tea	Tea	2006	164
	Cropland, other/no details	Cropland, no details	Cropland, no details	2006	168
Korea, Republic of	No information	No information	No information	2006	8'559
Kyr-gyzstan	Arable land	Textile fibers	Cotton	2006	140
	No information	No information	No information	2006	2'400
Latvia	Arable land	Cereals	Barley	2005	4'096
			Oats	2005	5'706
			Other cereals	2005	3'009
			Rye and meslin	2005	1'767
			Triticale	2005	766
			Wheat	2005	3'642
		Flowers and ornamental plants	Other Flowers and ornamental plants	2005	1
		Green fodder from arable land	Maize for silage	2005	24
			Temporary grassland	2005	63'730
		Industrial crops	Other Industrial crops	2005	26
		Medicinal & aromatic plants	Other Medicinal & aromatic plants	2005	27

Country	Main use	Crop category	Crop	Year	Organic agr. area (ha)
		Oilseeds	Linseed	2005	6
			Rape and Turnip rape	2005	1'155
		Protein crops	Dried pulses	2005	587
		Root crops	Fodder roots and brassicas	2005	254
			Potatoes	2005	5'358
			Sugar beets	2005	6
		Seeds and seedlings	Seeds and seedlings	2005	160
		Vegetables	Other vegetables	2005	214
	Other	Fallow land	Fallow land	2005	1'956
	Permanent crops	Fruit and nuts	Apples	2005	284
			Cherries	2005	29
			Fruit and nuts, no details	2005	551
			Pears	2005	19
			Plums	2005	16
	Permanent grassland	Pastures and meadows	Pastures and meadows	2005	25'223
Lebanon	Arable land	Cereals	Cereals, no details	2006	484
		Fallow land as part of crop rotation	Fallow land as part of crop rotation	2006	306
		Green fodder from arable land	Green fodder from arable land	2006	54
		Industrial crops	Other Industrial crops	2006	1'576
		Root crops	Other root crops	2006	20
		Vegetables	Other vegetables	2006	125
	No information	No information	No information	2006	-7
	Other	Fallow land	Fallow land	2006	22
	Permanent crops	Citrus fruit	Citrus fruit, no details	2006	18
		Fruit and nuts	Fruit and nuts, no details	2006	277
		Grapes	Grapes	2005	
				2006	188
		Olives	Olives	2006	79
	Permanent grassland	Permanent grassland, no details	Permanent grassland, no details	2006	328
Liechten-stein	Arable land	Cereals	Spelt	2006	5
			Wheat	2006	45
		Green fodder from arable land	Maize for silage	2006	70
			Temporary grassland	2006	150
		Oilseeds	Other oilseeds	2006	5
		Root crops	Potatoes	2006	10
			Root crops, no details	2006	2
		Vegetables	Vegetables, no details	2006	14
	No information	No information	No information	2005	36
	Permanent grassland	Pastures and meadows	Pastures and meadows	2005	690
Lithuania	Arable land	Cereals	Barley	2006	9'283
			Grain Maize	2006	150
			Oats	2006	7'054
			Other cereals	2006	13'479
			Rye and meslin	2006	7'402
			Wheat	2006	8'096
		Fallow land as part of crop rotation	Fallow land as part of crop rotation	2006	1'898
		Green fodder from arable land	Field fodder growing	2006	977
		Industrial crops	Textile fibers, no	2006	224

252

Country	Main use	Crop category	Crop	Year	Organic agr. area (ha)
			details		
		Medicinal & aromatic plants	Medicinal & aromatic plants	2006	558
		Oilseeds	Rape and Turnip rape	2006	2'334
		Protein crops	Dried curcuma	2006	17'153
		Root crops	Fodder roots and brassicas	2006	23
			Potatoes	2006	512
		Vegetables	Vegetables, no details	2006	165
	No information	No information	No information	2006	0
	Other	Fallow land	Fallow land	2006	429
	Permanent crops	Fruit and nuts	Apples	2006	1'122
			Berries, no details	2006	3'276
			Cherries	2006	2
			Fruit, no details	2006	22
			Nuts, no details	2006	28
			Pears	2006	33
			Plums	2006	28
	Permanent grassland	Permanent grassland, no details	Permanent grassland, no details	2006	22'470
Luxem-burg	Arable land	Cereals	Cereals, no details	2006	570
		Fallow land as part of crop rotation	Fallow land	2005	5
			Fallow land as part of crop rotation	2005	11
		Green fodder from arable land	Other field fodder growing	2005	549
		Industrial crops	Other Industrial crops	2005	2
		Oilseeds	Other oilseeds	2005	15
		Protein crops	Other protein crops	2005	67
		Root crops	Other root crops	2005	24
		Seeds and seedlings	Seeds and seedlings	2005	111
		Vegetables	Vegetables, no details	2005	23
	No information	No information	No information	2006	
	Permanent crops	Fruit and nuts	Fruit and nuts, no details	2006	45
		Grapes	Grapes	2005	6
				2006	
	Permanent grassland	Pastures and meadows	Pastures and meadows	2005	1'860
	Cropland, other/no details	Cropland, no details	Cropland, no details	2006	342
Macedo-nia	Arable land	Arable crops, no details	Arable crops, no details	2006	87
		Cereals	Barley	2006	75
			Grain Maize	2006	1
			Oats	2006	8
			Rye and meslin and meslin	2006	21
			Triticale	2006	11
			Wheat	2006	62
		Green fodder from arable land	Lucerne	2006	68
		Medicinal & aromatic plants	Lavender	2006	0.1
			Other Medicinal & aromatic plants	2006	2
			Rape and Turnip rape	2006	3
		Protein crops	Fodder pea	2006	8

Country	Main use	Crop category	Crop	Year	Organic agr. area (ha)
		Root crops	Potatoes	2006	18
		Vegetables	Beans	2006	34
			Tomatoes	2006	1
			Watermelon	2006	1
	Permanent crops	Fruit and nuts	Almonds	2006	28
			Apples	2006	0.2
			Apricots	2006	2
			Aronia	2006	0.2
			Blackberries	2006	1
			Cherries	2006	0.4
			Cherries, sour	2006	10
			Chestnuts	2006	7
			Hazelnut	2006	1
			Peaches	2006	1
			Pears	2006	0.1
			Plums	2006	4
			Pomegranate	2006	2
			Raspberries	2006	7
			Walnuts	2006	11
		Grapes	Grapes	2006	21
		Other permanent crops	Kaki	2006	12
			Kiwi	2006	1
Madagascar	No information	No information	No information	2006	9'456
Malawi	No information	No information	No information	2002	325
Malaysia	No information	No information	No information	2006	1'000
Mali	Arable land	Oilseeds	Sesame seeds	2006	457
		Textile fibers	Cotton	2006	1'663
	Permanent crops	Tropical fruit and nuts	Banana	2006	2
			Cashew nuts	2006	1
			Mango	2006	208
Malta	No information	No information	No information	2006	18
	Permanent crops	Grapes	Grapes	2006	2
Mauritius	No information	No information	No information	2006	
	Permanent crops	Sugarcane	Sugarcane	2006	175
Mexico	Arable land	Cereals	Amaranto	2005	193
			Grain Maize	2005	4'530
			Rice	2005	150
		Medicinal & aromatic plants	Aloe vera	2005	1'888
			Aromatic plants	2005	30'119
		Oilseeds	Cartamo	2005	662
			Sesame seeds	2005	2'498
		Other arable crops	Chile	2005	139
			Jamaica	2005	171
			Luffa	2005	36
		Protein crops	Frijol	2005	140
			Other protein crops	2005	156
		Vegetables	Other vegetables	2005	33'417
			Strawberries	2005	142
			Watermelon	2005	40
	No information	No information	No information	2005	13'789
				2006	96'426
	Permanent crops	Citrus fruit	Other citrus fruit	2005	1'608
		Cocoa	Cocoa	2005	17'314

254

Country	Main use	Crop category	Crop	Year	Organic agr. area (ha)
		Coffee	Coffee	2005	147'137
			Coffee associated with other crops	2005	2'906
		Fruit and nuts	Apples	2005	254
			Blackberries	2005	229
			Fruit and nuts, no details	2005	5'871
			Nuts, no details	2005	20
			Peaches	2005	8
			Pears	2005	4
			Plums	2005	5
			Raspberry	2005	263
		Olives	Olives	2005	1'000
		Other permanent crops	Neem	2005	213
			Vanilla	2005	571
			Yucca	2005	500
		Sugarcane	Sugarcane	2005	853
		Tropical fruit and nuts	Avocado	2005	2'652
			Banana	2005	153
			Cactus	2005	10'982
			Cashew nuts	2005	242
			Coconuts	2005	8'400
			Guava	2005	624
			Litchi	2005	104
			Macadamia	2005	28
			Mamey	2005	17
			Mango	2005	2'132
			Nanche	2005	15
			Papaya	2005	12
			Passion fruit	2005	4
			Pineapples	2005	253
			Pitaya	2005	15
	Permanent grassland	Permanent grassland, no details	Permanent grassland, no details	2005	15'233
Moldova	Arable land	Arable crops, no details	Arable crops, no details	2006	3'730
	No information	No information	No information	2006	720
	Permanent crops	Fruit and nuts	Fruit and nuts, no details	2006	450
			Walnuts	2006	2'178
		Grapes	Grapes	2006	4'327
Monte-negro	No information	No information	No information	2006	25'051
Morocco	Arable land	Flowers and ornamental plants	Roses	2006	30
		Medicinal & aromatic plants	Other Medicinal & aromatic plants	2006	1'726
		Vegetables	Other vegetables	2006	593
	No information	No information	No information	2006	
	Permanent crops	Fruit and nuts	Fruit, no details	2006	1'674
		Other permanent crops	Caper	2006	193
Mozam-bique	Arable land	Oilseeds	Peanuts	2006	482
			Sesame seeds	2006	246
	No information	No information	No information	2006	
Nepal	Arable land	Medicinal & aromatic plants	Herbs for essential oil	2006	6'646

255

Country	Main use	Crop category	Crop	Year	Organic agr. area (ha)
		Oilseeds	Sesame seeds	2006	50
	Permanent crops	Coffee	Coffee	2006	558
		Tea	Tea	2006	508
Netherlands	Arable land	Cereals	Cereals, no details	2006	5'168
		Fallow land as part of crop rotation	Fallow land as part of crop rotation	2006	1'070
		Green fodder from arable land	Field fodder growing	2006	2'834
		Vegetables	Vegetables, no details	2006	4'584
	No information	No information	No information	2006	
	Permanent crops	Fruit and nuts	Fruit, no details	2006	500
	Permanent grassland	Permanent grassland, no details	Permanent grassland, no details	2006	31'102
	Cropland, other/no details	Cropland, no details	Cropland, no details	2006	3'166
New Zealand	No information	No information	No information	2006	63'883
Nicaragua	No information	No information	No information	2005	5729
	Permanent grassland	Permanent grassland, no details	Permanent grassland, no details	2005	13'867
	Cropland, other/no details	Cropland, no details	Cropland, no details	2005	40'404
Niger	No information	No information	No information	2006	81
Nigeria	Arable land	Arable crops, no details	Arable crops, no details	2006	1
		Cereals	Rice	2006	8
		Other arable crops	Ginger	2006	6
			Maize, yam, amaranths, celosia	2006	1'000
			Yam and maize intercrop	2006	2'000
		Vegetables	Tomatoes	2006	1
			Vegetables, no details	2006	2
	No information	No information	No information	2006	
	Permanent crops	Citrus fruit	Lemon	2006	20
		Other permanent crops	Hibiscus	2006	4
Niue	No information	No information	No information	2006	159
Norway	Arable land	Arable crops, no details	Arable crops, no details	2006	918
		Cereals	Barley	2006	2'863
			Oats	2006	2'584
			Other cereals	2006	0
			Rye and meslin	2006	185
			Wheat	2006	1'077
		Fallow land as part of crop rotation	Fallow land as part of crop rotation	2006	105
		Flowers and ornamental plants	Flowers and ornamental plants	2006	2
		Green fodder from arable land	Green fodder from arable land, no details	2006	28'987
		Industrial crops	Rape and Turnip rape	2006	6
		Medicinal & aromatic plants	Medicinal & aromatic plants	2006	19
		Protein crops	Dried pulses	2006	179
		Root crops	Potatoes	2006	175
		Seeds and seedlings	Seeds and seedlings	2006	399
		Vegetables	Fresh Vegetables	2006	143
	No information	No information	No information	2006	
	Other	Fallow land	Fallow land	2006	99

256

Country	Main use	Crop category	Crop	Year	Organic agr. area (ha)
	Permanent crops	Fruit and nuts	Fruit and nuts, no details	2006	127
		Permanent crops, no details	Permanent crops, no details	2006	16
	Permanent grassland	Pastures and meadows	Pastures and meadows	2006	6'740
Pakistan	Arable land	Arable crops, no details	Arable crops, no details	2004	1'120
		Cereals	Rice	2004	6'360
			Wheat	2004	3'800
		Oilseeds	Sesame seeds	2004	3'560
		Textile fibers	Cotton	2004	880
	No information	No information	No information	2006	4'691
	Permanent crops	Citrus fruit	Other citrus fruit	2004	440
		Fruit and nuts	Other dry fruits	2004	1'600
		Tropical fruit and nuts	Other tropical fruits	2004	800
	Permanent grassland	Permanent grassland, no details	Permanent grassland, no details	2004	1'750
Palestine, Occupied Tr.	No information	No information	No information	2006	641
Panama	Arable land	Vegetables	Other vegetables	2005	7
	No information	No information	No information	2006	23
	Permanent crops	Citrus fruit	Other citrus fruit	2005	10
		Cocoa	Cocoa	2005	4'850
		Coffee	Coffee	2005	40
		Tropical fruit and nuts	Araza	2005	15
			Banana	2005	22
			Pineapples	2005	10
	Cropland, other/no details	Cropland, no details	Cropland, no details	2005	290
Papua New Guinea	No information	No information	No information	2006	2'497
Paraguay	Arable land	Medicinal & aromatic plants	Lemon verbena	2006	2
			Stevia	2006	8
		Oilseeds	Sesame seeds	2006	348
		Textile fibers	Cotton	2006	374
	Permanent crops	Fruit and nuts	oranges	2006	30
		Sugarcane	Sugarcane	2006	16'930
		Tea	Mate	2006	13
Peru	Arable land	Cereals	Andean grains (quinoa and kiwicha)	2006	1'996
			Grain Maize	2006	1'132
		Medicinal & aromatic plants	Aromatic plants	2006	2'165
		Oilseeds	Sacha inchi	2006	507
			Sesame seeds	2006	233
		Other arable crops	Maca	2006	483
			Panela	2006	658
			Yacon	2006	20
		Protein crops	Frijol	2006	122
			Other protein crops	2006	40
		Root crops	Potatoes	2006	36
		Textile fibers	Cotton	2006	2'597
		Vegetables	Tomatoes	2006	22
			Vegetables, no details	2006	1'157
	No information	No information	No information	2006	
	Permanent crops	Cocoa	Cocoa	2006	9'640

Country	Main use	Crop category	Crop	Year	Organic agr. area (ha)
		Coffee	Coffee	2006	72'095
		Fruit and nuts	Pecano	2006	126
		Grapes	Grapes	2006	20
		Olives	Olives	2006	145
		Other permanent crops	Jojoba	2006	27
		Sugarcane	Sugarcane	2006	68
		Tea	Tea	2006	136
		Tropical fruit and nuts	Avocado	2006	42
			Banana	2006	5'092
			Camu camu	2006	136
			Mango	2006	2'887
			Palmito	2006	20'000
			Passion fruit	2006	56
	Cropland, other/no details	Cropland, no details	Cropland, no details	2006	41
Philippines	No information	No information	No information	2006	5'691
Poland	Arable land	Arable crops, no details	Arable crops, no details	2006	91'203
		Vegetables	Vegetables, no details	2006	2'280
	No information	No information	No information	2006	2'281
	Permanent crops	Fruit and nuts	Fruit and nuts, no details	2006	50'162
	Permanent grassland	Permanent grassland, no details	Permanent grassland, no details	2006	82'083
Portugal	Arable land	Cereals	Other cereals	2006	41'588
		Fallow land as part of crop rotation	Fallow land as part of crop rotation	2006	1'278
		Medicinal & aromatic plants	Aromatic plants	2006	84
		Vegetables	Other vegetables	2006	883
	No information	No information	No information	2006	
	Other	Forest	Forest	2006	913
	Permanent crops	Fruit and nuts	Fruit and nuts, no details	2006	1'007
			Other dry fruits	2006	3'449
		Grapes	Grapes	2006	1'178
		Olives	Olives	2006	19'341
	Permanent grassland	Permanent grassland, no details	Permanent grassland, no details	2006	199'653
Romania	Arable land	Cereals	Barley	2006	1'278
			Grain Maize	2006	2'217
			Oats	2006	206
			Other cereals	2006	323
			Triticale	2006	321
			Wheat	2006	11'965
		Fallow land as part of crop rotation	Fallow land as part of crop rotation	2006	210
		Green fodder from arable land	Green fodder from arable land	2006	2'822
		Industrial crops	Textile fibers, no details	2006	114
		Medicinal & aromatic plants	Medicinal & aromatic plants	2006	1'761
		Oilseeds	Linseed	2006	68
			Other oilseeds	2006	0
			Rape and Turnip rape	2006	3'273
			Sunflower seed	2006	12'717

Country	Main use	Crop category	Crop	Year	Organic agr. area (ha)
		Protein crops	Dried pulses	2006	7'777
		Root crops	Other root crops	2006	0
			Potatoes	2006	15
			Sugar beets	2006	14
		Seeds and seedlings	Seeds and seedlings	2006	7
		Vegetables	Vegetables, no details	2006	727
	No information	No information	No information	2006	
	Permanent crops	Fruit and nuts	Apples	2006	108
			Berries, no details	2006	10
			Nuts, no details	2006	6
			Stone fruit	2006	87
		Grapes	Grapes	2006	83
	Permanent grassland	Pastures and meadows	Pastures and meadows	2006	49'807
		Rough Grazing	Rough Grazing	2006	1'393
	Cropland, other/no details	Cropland, no details	Cropland, no details	2006	10'273
Russian Federation, European Part	No information	No information	No information	2006	3'192
Rwanda	Arable land	Cereals	Grain Maize	2006	3
			Wheat	2006	2
		Medicinal & aromatic plants	Geranium (essential oil)	2006	4
		Protein crops	Beans	2006	4
		Root crops	Potatoes	2006	6
			Sweet potatoe	2006	2
		Vegetables	Chilies	2006	22
	No information	No information	No information	2006	4
	Permanent crops	Coffee	Coffee	2006	466
Samoa	No information	No information	No information	2006	7'243
Sao Tome and Prince	Arable land	Vegetables	Pepper	2006	14
	Permanent crops	Cocoa	Cocoa	2006	2'889
		Medicinal & aromatic plants	Vanilla	2006	14
Saudi Arabia	Arable land	Arable crops, no details	Arable crops, no details	2005	10'500
		Cereals	Barley	2005	25
			Wheat	2005	200
		Root crops	Potatoes	2005	10
		Vegetables	Greenhouse cultivation	2005	245
			Onion	2005	10
			Other vegetables	2005	186
			Tomatoes	2005	100
	Permanent crops	Fruit and nuts	Fruit and nuts, no details	2005	1'216
		Olives	Olives	2005	783
		Tropical fruit and nuts	Dates	2005	455
Senegal	Arable land	Cereals	Fonio	2006	38
			Grain Maize	2006	3
			Other cereals	2006	2
		Oilseeds	Sesame seeds	2006	37
		Textile fibers	Cotton	2006	51
	No information	No information	No information	2006	

Country	Main use	Crop category	Crop	Year	Organic agr. area (ha)
Serbia	Arable land	Cereals	Barley	2006	13
			Grain Maize	2006	55
			Oats	2006	4
			Rye and meslin	2006	1
			Triticale	2006	8
			Wheat	2006	259
		Green fodder from arable land	Field fodder growing	2006	31
		Medicinal & aromatic plants	Other Medicinal & aromatic plants	2006	3
		Oilseeds	Flax/linseed	2006	1
			Soy	2006	25
			Sunflower seed	2006	214
		Root crops	Other root crops	2006	5
			Podded peas	2006	1
		Vegetables	Vegetables, no details	2006	137
	No information	No information	No information	2006	
	Permanent crops	Fruit and nuts	Apples	2006	15
			Fruit and nuts, no details	2006	58
			Raspberries	2006	71
		Grapes	Grapes	2006	6
Slovak Republic	Arable land	Arable crops, no details	Arable crops, no details	2006	22'231
		Cereals	Other cereals	2006	15'000
	Permanent crops	Fruit and nuts	Fruit, no details	2006	679
		Grapes	Grapes	2006	53
	Permanent grassland	Permanent grassland, no details	Permanent grassland, no details	2006	83'498
Slovenia	Arable land	Cereals	Barley	2006	126
			Grain Maize	2006	132
			Other cereals	2006	284
			Rye and meslin	2006	66
			Triticale	2006	141
			Wheat	2006	140
		Fallow land as part of crop rotation	Fallow land as part of crop rotation	2006	49
		Green fodder from arable land	Green fodder from arable land	2006	421
		Industrial crops	Other Industrial crops	2006	20
		Medicinal & aromatic plants	Medicinal & aromatic plants	2006	9
		Oilseeds	Other oilseeds	2006	36
			Soy	2006	5
			Sunflower seed	2006	10
		Protein crops	Dried pulses	2006	63
		Root crops	Other root crops	2006	2
			Potatoes	2006	83
		Vegetables	Vegetables, no details	2006	96
	No information	No information	No information	2006	2
	Permanent crops	Fruit and nuts	Fruit and nuts, no details	2006	536
		Grapes	Grapes	2006	125
		Olives	Olives	2006	27
	Permanent grassland	Permanent grassland, no details	Permanent grassland, no details	2006	24'458
Solomon Islands	No information	No information	No information	2006	3'628

Country	Main use	Crop category	Crop	Year	Organic agr. area (ha)
South Africa	No information	No information	No information	2005	50'000
Spain	Arable land	Cereals	Cereals, leguminous crops, other	2006	113'304
		Fallow land as part of crop rotation	Fallow land as part of crop rotation	2006	55'159
		Medicinal & aromatic plants	Other Medicinal & aromatic plants	2006	15'051
		Seeds and seedlings	Seeds and seedlings	2006	7'420
		Vegetables	Other vegetables	2006	5'039
	No information	No information	No information	2006	
	Other	Forest	Forest	2006	189'452
	Permanent crops	Citrus fruit	Other citrus fruit	2006	2'184
		Fruit and nuts	Fruit and nuts, no details	2006	3'866
			Other dry fruits	2006	44'600
		Grapes	Grapes	2006	16'832
		Olives	Olives	2006	93'432
		Tropical fruit and nuts	Banana associated with other crops	2006	773
	Permanent grassland	Permanent grassland, no details	Permanent grassland, no details	2006	378'820
	Cropland, other/no details	Cropland, no details	Cropland, no details	2006	458
Sri Lanka	Arable land	Medicinal & aromatic plants	Herbs	2006	30
			Spices	2006	300
	No information	No information	No information	2006	11'637
	Permanent crops	Fruit and nuts	Fruit, no details	2006	1'814
		Tea	Black tea	2006	2'280
			Green tea	2006	490
		Tropical fruit and nuts	Cashew nuts	2006	142
			Coconuts	2006	307
Sudan	No information	No information	No information	2006	
Sweden	Arable land	Arable crops, no details	Arable crops, no details	2006	6'921
		Cereals	Barley	2006	16'730
			Oats	2006	31'240
			Rye and meslin	2006	2'980
			Triticale	2006	4'710
			Wheat, summer	2006	7'290
			Wheat, winter	2006	12'620
		Fallow land as part of crop rotation	Fallow land as part of crop rotation	2006	559
		Green fodder from arable land	Green fodder from arable land	2006	77'563
		Industrial crops	Industrial crops	2006	229
		Oilseeds	Linseed	2006	470
			Rape and Turnip rape	2006	4'330
		Protein crops	Fodder pea	2006	4'200
			Fodder beans	2006	4'150
		Root crops	Potatoes	2006	870
		Seeds and seedlings	Seeds and seedlings	2006	4'180
		Textile fibers	Textile fibers, no details	2006	137
		Vegetables	Other vegetables	2006	561
	No information	No information	No information	2006	342
	Permanent crops	Fruit and nuts	Fruit, no details	2006	287
	Permanent grassland	Permanent grassland,	Permanent grassland,	2006	40'026

Country	Main use	Crop category	Crop	Year	Organic agr. area (ha)
		no details	no details		
	Cropland, other/no details	Cropland, no details	Cropland, no details	2006	4'990
Switzerland	Arable land	Arable crops, no details	Arable crops, no details	2006	1'412
		Cereals	Barley	2005	791
			Cereals, no details	2006	27
			Grain Maize	2005	401
			Oats	2005	187
			Other cereals	2005	17
			Rye and meslin and meslin	2005	242
			Spelt	2005	761
			Triticale	2005	308
			Wheat	2005	2'327
			Wheat, emmer	2005	39
		Fallow land as part of crop rotation	Fallow land as part of crop rotation	2005	134
		Flowers and ornamental plants	Other Flowers and ornamental plants	2005	21
		Green fodder from arable land	Feed legumes	2005	195
			Maize for silage	2005	1'236
			Temporary grassland	2005	8'367
		Industrial crops	Tobacco	2005	2
		Medicinal & aromatic plants	Other Medicinal & aromatic plants	2005	26
		Oilseeds	Oilseeds, no details	2006	54
			Pumpkin seeds	2005	13
			Rape seeds	2005	56
			Soy	2005	38
			Sunflower seed	2005	39
		Root crops	Fodder beet	2005	16
			Potatoes	2005	481
			Root crops, no details	2006	93
			Sugar beets	2005	10
		Textile fibers	Flax	2005	8
		Vegetables	Brussels sprouts	2005	3
			Greenhouse cultivation	2005	58
			Other vegetables	2005	1'136
	No information	No information	No information	2006	8'479
	Permanent crops	Fruit and nuts	Apples	2005	367
			Fruit, no details	2005	89
			Pears	2005	62
		Grapes	Grapes	2005	388
				2006	
		Industrial crops	Hops	2005	3
		Medicinal & aromatic plants	Other Medicinal & aromatic plants	2005	88
			Rhubarb	2005	8
		Permanent crops, no details	Permanent crops, no details	2005	1
				2006	482
	Permanent grassland	Pastures and meadows	Pastures and meadows	2005	78'588
		Permanent grassland, no details	Permanent grassland, no details	2005	18'543
Syria	Arable land	Textile fibers	Cotton	2006	27'881

Country	Main use	Crop category	Crop	Year	Organic agr. area (ha)
	No information	No information	No information	2006	2'032
	Permanent crops	Grapes	Grapes	2006	10
		Olives	Olives	2006	570
Taiwan	Arable land	Cereals	Grain Maize	2006	1
			Rice	2006	720
		Flowers and ornamental plants	Other Flowers and ornamental plants	2006	2
		Medicinal & aromatic plants	Other Medicinal & aromatic plants	2006	5
		Root crops	Other root crops	2006	14
		Vegetables	Other vegetables	2006	101
			Strawberries	2006	2
			Tomatoes	2006	1
	No information	No information	No information	2006	459
	Permanent crops	Citrus fruit	oranges	2006	8
		Coffee	Coffee	2006	0
		Fruit and nuts	Fruit and nuts, no details	2006	16
			Fruit, no details	2006	4
			Peaches	2006	8
		Grapes	Grapes	2006	1
		Tea	Tea	2006	60
		Tropical fruit and nuts	Banana	2006	2
			Cactus	2006	6
			Guava	2006	1
			Litchi	2006	3
			Mango	2006	0
			Other tropical fruits	2006	11
			Pineapples	2006	5
	Permanent grassland	Permanent grassland, no details	Permanent grassland, no details	2006	313
	Cropland, other/no details	Mushrooms	Mushrooms	2006	2
Tanzania	Arable land	Medicinal & aromatic plants	Essential oil	2006	50
		Oilseeds	Peanuts	2006	4'404
			Sesame seeds	2006	1'800
		Textile fibers	Cotton	2006	5'748
	No information	No information	No information	2006	
	Permanent crops	Coffee	Coffee, arabica	2006	3'057
			Robusta coffee	2006	1'525
		Fruit and nuts	Dried fruit	2006	294
		Medicinal & aromatic plants	Spices	2006	5'488
		Tropical fruit and nuts	Cashew nuts	2006	1'286
			Mango	2006	80
Thailand	Arable land	Cereals	Other cereals	2006	1'077
			Rice	2006	17'328
		Vegetables	Other vegetables	2006	2'375
	No information	No information	No information	2006	
	Permanent crops	Fruit and nuts	Fruit and nuts, no details	2006	799
	Cropland, other/no details	Cropland, no details	Cropland, no details	2006	122
Timor Leste	No information	No information	No information	2006	2'063
	Permanent crops	Coffee	Coffee	2005	21'325
		Other permanent	cloves	2005	71

263

Country	Main use	Crop category	Crop	Year	Organic agr. area (ha)
		crops			
			Vanilla	2005	130
Togo	Arable land	Oilseeds	Soy	2006	1'501
	No information	No information	No information	2006	73
	Permanent crops	Tropical fruit and nuts	Avocado	2006	71
			Banana	2006	15
			Cashew nuts	2006	286
			Mango	2006	69
			Papaya	2006	2
			Pineapples	2006	321
Trinidad & Tobago	No information	No information	No information	2005	67
Tunisia	Arable land	Medicinal & aromatic plants	Medicinal & aromatic plants	2006	5'854
		Other arable crops	Cereals and forage	2006	1'746
		Vegetables	Vegetables, no details	2006	63
	No information	No information	No information	2006	
	Permanent crops	Fruit and nuts	Almonds	2006	4'368
			Fruit trees and cactus	2006	2'062
		Olives	Olives	2006	89'324
		Tropical fruit and nuts	Dates	2006	1'071
	Permanent grassland	Permanent grassland, no details	Permanent grassland, no details	2006	50'305
Turkey	Arable land	Arable crops, no details	Arable crops, no details	2006	-48'643[1]
		Cereals	Barley	2006	2'391
			Grain Maize	2006	1'006
			Oats	2006	121
			Rice	2006	77
			Rye and meslin and meslin	2006	402
			Wheat and spelt	2006	14'955
		Fallow land as part of crop rotation	Fallow land as part of crop rotation	2006	9'152
		Green fodder from arable land	Maize for silage	2006	107
			Other field fodder growing	2006	1'480
		Industrial crops	Other Industrial crops	2006	1'103
			Soy	2006	4
			Sunflower seed	2006	137
		Medicinal & aromatic plants	Medicinal & aromatic plants	2006	0
		Protein crops	Dried pulses	2006	7'637
		Root crops	Fodder beet	2006	1
			Potatoes	2006	107
			Sugar beets	2006	14
		Textile fibers	Cotton	2006	16'641
		Vegetables	Vegetables, no details	2006	2'879
	No information	No information	No information	2006	
	Permanent crops	Citrus fruit	Grapefruit	2006	67
			Lemon	2006	132
			oranges	2006	311
			Other citrus fruit	2006	164
		Fruit and nuts	Apples	2006	4'278

---

[1] In Turkey, most of the arable land is used for two main crops. In order to be able to list all crops but still have the correct total, this correction value is needed.

Country	Main use	Crop category	Crop	Year	Organic agr. area (ha)
			Apricots	2006	3'583
			Cherries	2006	573
			Fruit, no details	2006	38'643
			Nuts, no details	2006	13'438
			Other stone fruit	2006	1'990
			Peaches	2006	151
			Pears	2006	2'627
			Plums	2006	2'471
		Grapes	Grapes	2006	5'485
		Olives	Olives	2006	8'272
		Permanent crops, no details	Permanent crops, no details	2006	587
	Permanent grassland	Pastures and meadows	Pastures and meadows	2006	268
		Rough Grazing	Rough Grazing	2006	7'664
Uganda	No information	No information	No information	2006	66'968
	Permanent crops	Cocoa	Cocoa	2006	3'750
		Coffee	Coffee	2006	17'721
UK	Arable land	Arable crops, no details	Arable crops, no details	2006	3'802
		Cereals	Barley	2006	11'661
			Grain Maize	2006	762
			Oats	2006	8'051
			Other cereals	2006	5'268
			Wheat	2006	21'782
		Fallow land as part of crop rotation	Fallow land as part of crop rotation	2006	2'408
		Flowers and ornamental plants	Flowers and ornamental plants	2006	23
		Green fodder from arable land	Green fodder from arable land, no details	2006	350
			Temporary grassland	2006	101'950
		Industrial crops	Other industrial crops	2006	1'929
			Rape and Turnip rape	2006	679
		Medicinal & aromatic plants	Medicinal & aromatic plants	2006	670
		Protein crops	Dried pulses	2006	156
		Root crops	Fodder roots and brassicas	2006	22
			Potatoes	2006	2'360
			Sugar beets	2006	267
		Seeds and seedlings	Seeds and seedlings	2006	39
		Vegetables	Vegetables, no details	2006	11'287
	No information	No information	No information	2006	-188
	Other	Fallow land	Fallow land	2006	3'072
	Permanent crops	Fruit and nuts	Fruit and nuts, no details	2006	1'781
		Grapes	Grapes	2006	29
		Permanent crops, no details	Permanent crops, no details	2006	3'082
	Permanent grassland	Pastures and meadows	Pastures and meadows	2006	383'002
		Rough Grazing	Rough Grazing	2006	40'327
Ukraine	Arable land	Cereals	Oats	2006	2'210
			Other cereals	2006	149'288
			Rye and meslin	2006	850
			Sorghum	2006	30
			Spelt	2006	2
			Triticale	2006	240
			Wheat, durum	2006	380

Country	Main use	Crop category	Crop	Year	Organic agr. area (ha)
		Fallow land as part of crop rotation	Fallow land as part of crop rotation	2006	14'100
		Flowers and ornamental plants	Roses	2006	28
		Green fodder from arable land	Green fodder from arable land, no details	2006	9'420
			Lucerne	2006	9'300
			Maize for silage	2006	6'000
			Temporary grassland	2006	16'880
		Medicinal & aromatic plants	Chamomile	2006	46
			Herbs for essential oil	2006	1'100
			Lavender	2006	650
			Other Medicinal & aromatic plants	2006	45
		Oilseeds	Rape and Turnip rape	2006	2'980
			Soy	2006	330
			Sunflower seed	2006	12'500
		Protein crops	Other protein crops	2006	1'600
			Peas	2006	4'700
			Soy	2006	350
		Root crops	Other root crops	2006	150
			Potatoes	2006	50
			Sugar beets	2006	3'200
		Textile fibers	Flax	2006	20
		Vegetables	Strawberries	2006	2
			Vegetables, no details	2006	40
	No information	No information	No information	2006	-2'369
	Other	Fallow land	Fallow land	2006	25'400
	Permanent crops	Fruit and nuts	Apples	2006	400
			Apricots	2006	4
			Fruit and nuts, no details	2006	5
			Nuts, no details	2006	100
			Pears	2006	2
			Plums	2006	1
Uruguay	Arable land	Cereals	Cereals, no details	2006	2'800
		Medicinal & aromatic plants	Medicinal & aromatic plants	2006	16
		Oilseeds	Sunflower seed	2006	200
		Vegetables	Vegetables, no details	2006	300
	No information	No information	No information	2006	
	Permanent crops	Citrus fruit	Citrus fruit, no details	2006	410
		Grapes	Grapes	2006	40
		Olives	Olives	2006	425
	Permanent grassland	Permanent grassland, no details	Permanent grassland, no details	2006	300
		Rough Grazing	Rough Grazing	2006	926'474
USA	Arable land	Cereals	Barley	2005	15'892
			Buckwheat	2005	2'575
			Grain Maize	2005	52'881
			Millet	2005	5'736
			Oats	2005	18'804
			Rice	2005	10'695
			Rye and meslin and meslin	2005	3'479
			Sorghum	2005	2'445

Country	Main use	Crop category	Crop	Year	Organic agr. area (ha)
			Spelt	2005	3'306
			Wheat	2005	112'295
		Fallow land as part of crop rotation	Fallow land as part of crop rotation	2005	66'789
		Green fodder from arable land	Other field fodder growing	2005	166'464
		Medicinal & aromatic plants	Other Medicinal & aromatic plants	2005	2'125
		Oilseeds	Peanuts	2005	4'832
			Sunflower seed	2005	2'463
		Protein crops	Beans	2005	4'274
			Dry peas and lentils	2005	7'186
			Soy	2005	49'459
		Root crops	Potatoes	2005	2'663
		Textile fibers	Cotton	2005	3'859
			Flax	2005	12'482
		Vegetables	Carrot	2005	2'322
			lettuce	2005	4'851
			Other vegetables	2005	29'765
			Tomatoes	2005	2'693
	No information	No information	No information	2005	62'483
	Permanent crops	Citrus fruit	Other citrus fruit	2005	4'107
		Fruit and nuts	Apples	2005	5'168
			Fruit and nuts, no details	2005	14'368
			Nuts, no details	2005	6'469
		Grapes	Grapes	2005	9'209
		Other permanent crops	Trees for maple syrup	2005	4'956
	Permanent grassland	Pastures and meadows	Pastures and meadows	2005	923'253
Vanuatu	No information	No information	No information	2006	8'996
Venezuela	No information	No information	No information	2006	15'712
Vietnam	No information	No information	No information	2006	21'867
Zambia	Arable land	Medicinal & aromatic plants	Medicinal & aromatic plants	2006	150
		Oilseeds	Oilseeds	2006	100
		Vegetables	Vegetables, no details	2006	864
	No information	No information	No information	2006	
	Cropland, other/no details	Cropland, no details	Cropland, no details	2006	1'253
Total					30'418'261

For Product Safety Concerns and Information please contact our EU representative GPSR@taylorandfrancis.com Taylor & Francis Verlag GmbH, Kaufingerstraße 24, 80331 München, Germany

Batch number: 08153780

Printed by Printforce, the Netherlands